RAPID THERMAL PROCESSING

RAPID THERMAL PROCESSING

PROCEEDINGS OF SYMPOSIUM I ON
RAPID THERMAL PROCESSING
OF THE E-MRS 1998 SPRING CONFERENCE

STRASBOURG, FRANCE, 16-19 JUNE 1998

Edited by

A. SLAOUI
CNRS/PHASE, Strasbourg, France

R.K. SINGH
University of Florida, USA

T. THEILER
STEAG-AST, Dornstadt, Germany

J.C. MULLER
CNRS/PHASE, Strasbourg, France

ELSEVIER
AMSTERDAM - LAUSANNE - NEW YORK - OXFORD - SHANNON - SINGAPORE - TOKYO

ELSEVIER SCIENCE Ltd
The Boulevard, Langford Lane
Kidlington, Oxford OX5 1GB, UK

ISBN: 9780080436128

Reprinted from: Materials Science in Semiconductor Processing, vol. 1/3-4

Transferred to digital print 2009
Printed and bound in Great Britain by CPI Antony Rowe, Chippenham and Eastbourne

MATERIALS SCIENCE IN SEMICONDUCTOR PROCESSING

PERGAMON

Materials Science in Semiconductor Processing 1 (1998) 167

MATERIALS
SCIENCE IN
SEMICONDUCTOR
PROCESSING

Preface

Rapid Thermal Processing (RTP) is a well established single-wafer technology in USLI semiconductor manufacturing and electrical engineering, as well as in materials science. The biggest advantage of RTP is that it eliminates the long-ramp-up and ramp-down times associated with furnaces, enabling a significant reduction in the thermal budget. Today, RTP is in production use for source/drain implant annealing, contact alloying, formation of refractory nitrides and silicides and thin gate dielectric (oxide) formation. The aim of symposium I was to provide an overview of the latest information on research and development in the different topics cited above. The potential applications of RTP in new area like large area devices such as flat planel displays and solar cells has to be investigated.

About 30 papers were presented in this symposium. The contributions of most interest involved modelling and control, junctions formation and thermal oxidation, deposition and recrystallisation and silicide formations. However, the range of topics and the intent to focus on underlying, fundamental issues common to semiconductor materials can be appreciated by the following list of presentations on processing issues like dopant diffusion in silicon from solid sources, strain relaxation and photonic effects, nucleation as well as applications to magnetic films and solar cells devices.

I would like to take this opportunity to thank the European Material Research Society (EMRS) for the organization of this symposium. Also, many thanks to the co-chairmen, namely, Pr. R. Singh, (University of Gainesville, USA), Dr T. Theiler (STEAG-AST, Donstadt, D) and J. C. Muller (Lab. PHASE/CNRS, Strasbourg, Fr). Finally, this symposium has been financially supported by JIPELIC (Fr), AG Associate (USA) and STEAG-AST (D).

PERGAMON

Materials Science in Semiconductor Processing 1 (1998) 169–179

MATERIALS
SCIENCE IN
SEMICONDUCTOR
PROCESSING

Rapid thermal processing technology for the 21st century

P.J. Timans *

AG Associates, 4425 Fortran Drive, San Jose, CA 95134-2300, USA

Abstract

Rapid thermal processing (RTP) has emerged as a key manufacturing technique for semiconductor device fabrication. In comparison to conventional furnace processing, where large batches of wafers are loaded into tube furnaces, RTP presents advantages in temperature control, ambient purity, cycle time and process flexibility. The fundamental flexibility of RTP in creating new types of thermal processes arises from the dynamic control of the heat source temperature, which permits fast heating combined with dynamic optimization of temperature uniformity. This paper illustrates the evolution of RTP system design to include concepts such as wafer rotation, axisymmetric heater design and multiple point dynamic temperature control. The resulting improvements in temperature uniformity and repeatability are demonstrated using process results which show that the new generation of RTP equipment can control the temperature distribution on both 200 and 300 mm wafers with a 3σ (3 standard deviations) range below 3°C. The power requirements for fast heating processes where minimization of diffusion is essential, such as ion-implantation damage annealing, are analyzed, and it is shown that heating rates up to 280°C/s can be achieved on 300 mm wafers. RTP, as a uniquely agile manufacturing process, will greatly assist the introduction of the new materials and processes required to extend metal-oxide-semiconductor (MOS) technology to its natural limits over the next 20 years, and is also posed to make significant contributions in emerging technologies including nanoelectronics, microelectromechanical devices and nonsilicon applications, including the processing of magnetic materials. © 1999 Elsevier Science Ltd. All rights reserved.

1. Introduction

The continuous technological advance of semiconductor device manufacturing technology enables the development of increasingly sophisticated circuits for computing, data storage and communications. The advances are mainly achieved through reductions in the dimensions of the basic circuit building block, the MOS transistor and the corresponding scaling of line widths, layer thicknesses and doping densities. The Semiconductor Industry Association has recently published its latest roadmap, which presents projections of the expected device and manufacturing process requirements covering the next 15 years [1]. The MOS-style transistor will continue to provide the most important electronic element for the near future, but the end of

the ability to scale down devices using conventional MOS technology is approaching. Several key device fabrication methods will not be applicable when device dimensions shrink below 0.1 μm. Lithography for features < 0.1 μm poses a serious economic problem, but a more fundamental problem arises when the gate oxide thickness drops below ~15 Å, because of the large leakage currents associated with direct tunneling through such a thin oxide. The introduction of a new gate insulator material is an immense change in the technology, posing considerable problems in materials science and in manufacturing technology. The gate electrode material may also have to change from doped polysilicon to a metallic structure, in order to prevent carrier depletion effects. The tight control of doping profiles throughout the device also poses significant problems, including the formation of extremely shallow junctions. In order to prevent interconnect delays from limiting the circuit speed, the move to

* Tel.: + 1-408-935-2235; fax: + 1-408-935-2775.

lower resistivity interconnects, made of copper and low dielectric constant materials for interlayer dielectrics has already started. As device technology advances, the control of contamination and defects becomes increasingly critical for a high yield of working devices, which is essential to make the manufacturing process economical. The scale-down of devices will be accompanied by increases in wafer diameters, initially with the move to 300 mm wafers, which is expected to well underway by the year 2000, and then the introduction of 450 and 675 mm wafers in the years 2010 and 2020, respectively. These technological changes are accompanied by the evolution of automation in fabrication and increasingly sophisticated approaches to control of manufacturing processes.

From this brief survey, it is evident that the landscape of silicon device manufacturing is changing, and that these changes require new materials and manufacturing processes. Many of the manufacturing steps require thermal processing, for the deposition and growth of films, induction of reactions and phase changes, annealing of defects introduced by energetic processes such as ion-implantation and plasma processing and the control of stress and topography. The general trend in thermal processing is to reduce the process temperature and duration as much as possible in order to restrict the motion of atoms through atomic diffusion, which tends to change the effective shape of device structures and can cause undesirable side-reactions [2]. Alternative technologies to thermal processing, including the wider application of plasma-based techniques for film deposition, are developing in order to limit the thermal exposure, but even with these techniques it is often essential to provide suitable thermal anneals to permit the atomic bond rearrangements which reduce defect densities and stress [3, 4].

2. Thermal processing technology

At present, many thermal processes are performed using conventional hot-wall furnaces, which heat large batches of wafers to high temperatures slowly. Rapid thermal processing (RTP), which involves individually heating each wafer to a high temperature rapidly, is now widely used when the restriction of diffusion is especially important, for example in ion-implantation damage annealing, and when the control of impurities in the processing ambient is critical, for example in the formation of metal silicides [2, 5–7]. RTP technology represents a significant change from the batch furnace approach, and its acceptance into widespread use in semiconductor device manufacturing has been slow, mainly as a result of the inherently conservative approach to process development adopted by the industry, but also because technological problems

made early RTP systems less easy to use than furnaces [2, 8]. This situation has changed to the point that it is now reasonable to claim that RTP systems are simpler, more versatile, and more cost-effective than batch processing methods for many applications. This paper will demonstrate that RTP is likely become the dominant thermal processing technology for semiconductor device technology as we move into the 21st century when MOS-device technology will reach its natural limits and alternative device technologies will take over.

The most basic differences between RTP and furnaces arise from differences in the fundamental philosophy of thermal control. One essential requirement of any thermal processing system is to make all the points on the surface of every wafer processed experience the same temperature–time cycle as defined in the process recipe. If this requirement is met, then the thermal process is perfectly uniform and there is no differential thermal expansion within the wafer, and no thermal stress. Process nonuniformity across a wafer and poor repeatability between wafers can reduce device yield. Thermal stress is a serious problem if it generates defects which can degrade devices or cause crystallographic slip [9].

There are two basic approaches to thermal control. One approach is to arrange for the heat-transfer conditions in the system to be such that the thermal cycle at any point on any wafer is identical. This approach can be called open-loop control, and it represents the operation of a furnace, where there is no feedback into the control system about the temperature on any wafer or the temperature distribution across any wafer. Nevertheless, in many processes, furnaces are capable of excellent repeatability and temperature uniformity. This is because the cycle time for many processes is long, and the batch of wafers inevitably reaches thermal equilibrium with the furnace walls. Under these conditions the temperature uniformity and repeatability can be excellent if the furnace environment is kept sufficiently isothermal through control of the wall temperature and measures to deal with end-effects. As device technology advances and the exposure times at high temperature reduce to decrease the thermal exposure and to create thinner films, as well as to reduce the cycle time, the conventional approach often ceases to be adequate.

In contrast to the furnace environment, a wafer in an RTP system is never in true thermal equilibrium with the heat source. The heat source can be a bank of tungsten-halogen lamps, a hot-plate, or even a hot-wall environment, but in all cases the central feature is that the power source is significantly hotter than the wafer at all stages of the thermal cycle. This is absolutely necessary to allow rapid heating. The point can be illustrated through a simple calculation of the response

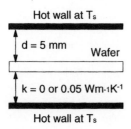

Fig. 1. Model for wafer heating in a hot-wall environment. The wafer is 5 mm from the wall surfaces and the ambient is either vacuum, when the thermal conductivity, k, is zero, or it is a gas with $k = 0.05$ W m^{-1} K^{-1}.

of a wafer when it is suddenly exposed to radiation from its surroundings as illustrated in Fig. 1. The calculations assume that the wafer and the heat sources are infinite in lateral extent and that the wafer is 725 μm thick, as is typical for a 200 mm wafer. The heat sources and the silicon wafer are assumed to be opaque and spectrally grey, with total emissivities of 0.8 and 0.68, respectively. The transient responses of the wafer shown in Fig. 2 were calculated for various heat source temperatures using methods described previously [10, 11]. Fig. 2 shows the difference in behavior when the heat transfer occurs purely by radiation, as might be expected in low-pressure (< 1 Torr) systems, and that when thermal conduction occurs. For the results shown, where the thermal conductivity of the gas was set at 0.05 Wm^{-1}K^{-1} and the wafer was assumed to be 5 mm away from the hot surfaces, conduction does not play a large part in the behavior. The heating rate rises rapidly with the source temperature and in order to achieve the highest heating rates the source temperature has to be set at a value well above the target temperature, which means that it is necessary to control the source temperature dynamically. Fig. 3 shows the relationships between the average ramp rate, deduced from the time it takes to heat up

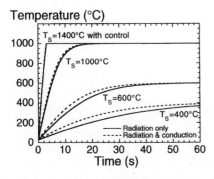

Fig. 2. Predictions of the transient response of a wafer suddenly exposed to a hot-wall environment. The curves show the responses for various wall temperatures, and compare the behavior in a vacuum with that when thermal conduction occurs under the conditions shown in Fig. 1.

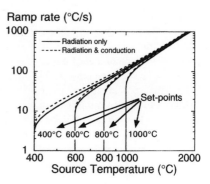

Fig. 3. The effect of the heat source temperature on the average ramp rate from 27°C to various recipe set-point temperatures. The behavior is shown for the cases where heat transfer occurs only by radiation and when conduction is included, for the conditions shown in Fig. 1.

to a given recipe temperature and the heat source temperature. The graph illustrates the need for dynamic control of the heat source temperature to achieve a desired ramp rate. For example, to reach 1000°C in 10 s and achieve a average ramp rate of 100°C/s, it is necessary for the heat source to be at ~1075°C during the ramp, and then to return to 1000°C in the steady state. For a recipe at 600°C, even a fairly modest ramp rate of 10°C/s calls for dynamic control of the source temperature. For a RTP system employing tungsten-halogen lamps, the source temperature is the temperature of the lamp filaments, which is easily adjusted dynamically and can be as high as ~2700°C. This is one of the basic strengths of RTP technology, because it allows enormous flexibility in the creation of processing recipes, including rapid heating to low temperatures, which is not available in conventional furnaces. Because the temperatures of the chamber and the heat source are not directly related to that of the wafer, RTP systems use real-time wafer temperature measurement to provide feedback for closed-loop control. The need to measure the temperature of a wafer to within a few °C posed a serious problem for early RTP technology [8]. In particular, pyrometric methods can be influenced by stray radiation from heat sources other than the wafer and by coatings which change the spectral emissivity of the wafer at the pyrometer wavelength [11, 12]. However, it has been found that simple, reliable and cost-effective temperature control strategies, in particular the use of radiation shields, can provide highly repeatable thermal cycles with excellent emissivity-independence [10, 11, 13]. Pyrometric methods which address the wafer emissivity problem have also developed [14, 15].

In a conventional furnace, the heat transfer conditions do not resemble those shown in Fig. 1. Generally, a stack of wafers is inserted into a hot tube which is then heated to the process temperature.

Wafers cannot heat up uniformly in this geometry, because their edges are exposed to the high temperature of the furnace wall and heat up faster than their centers [16]. This causes a radial temperature gradient to arise as the wafers are loaded into the furnace or as the wall temperature is increased to the process temperature. Likewise, during cooling, temperature gradients arise because the edges lose heat faster than the center. These temperature gradients have to be limited to prevent excessive thermal stress from causing slip. The temperature feedback is achieved from thermocouples which sense the temperatures at various positions in the furnace, but do not actually read the wafer temperature. At low heating rates, $\sim 5°C/min$, the pseudo-isothermal conditions mean that the wafer and the thermocouples are at similar temperatures. The small-batch, fast-ramp furnace has been introduced recently to try to overcome some of the limitations of conventional batch processing [17]. These systems rely on sophisticated model-based control schemes to maximize heating and cooling rates while preventing excessive stress [17, 18]. They process fewer wafers in one batch in order to increase the spacing between the wafers to allow faster heat transfer to the center of the wafer during temperature ramps [17]. As wafer diameters increase, the spacing between them also has to increase, so this problem of decreasing batch size becomes worse. These furnaces are a clear move in the direction of RTP, because the heater temperature is different to the wafer temperature during temperature ramps, the control requires dynamic estimation of the wafer temperature, there are fewer wafers in the batch and the cycle time is shorter. RTP systems address dynamic temperature control through a heating system which allows dynamic uniformity tuning, freeing the system from the inherent constraints which apply to furnace processing. As a result process uniformity can exceed that attainable in furnaces and the scale up of the systems to larger wafer diameters does not pose a fundamental problem.

The advantages of RTP relative to batch processing do not end with thermal control. One of the main factors which resulted in the widespread deployment of RTP systems in semiconductor device manufacturing was the ability to control the purity of the gas ambient around the wafer to a much better level than was possible in furnaces [7]. The high purity ambient and the ability to rapidly switch between different process gases for sophisticated processes stems from the small chamber volume of a single-wafer processing system. Good ambient purity control, combined with low chamber wall temperatures and short process times, give RTP systems a strong advantage over furnaces in minimizing wafer contamination. RTP systems are also inherently compatible with in situ monitors, including in-chamber gas purity sensors, which can provide protection against contamination for every wafer processed, reducing the risks and improving yield. Another potential strength of RTP will become clear if photo-enhanced processes prove to be important, because conventional furnaces cannot exploit these methods because their geometry is incompatible with uniform illumination of the wafers [19].

Cost-of-ownership models can be used to compare the cost per wafer of running a given process in batch furnaces or RTP systems [20]. These models account for many factors including the system price, throughput, the clean-room space required, reliability and running costs. This is a difficult problem to analyze correctly, and the results depend on assumptions which should be tailored differently for each device manufacturer. Typically RTP systems emerge as being superior when the process recipe calls for a short time at high temperature, while the batch systems win out for longer processes, where the long loading, ramp-up, ramp-down and unloading times of a batch system have less impact on the throughput. Several trends suggest that RTP may soon become more economical than batch processing for most processes. One trend arises from the scaling-down of device dimensions which reduces the film thicknesses in devices, favoring shorter growth and deposition times [5]. Another trend is that of increasing wafer diameter, which decreases the number of wafers which can be heated uniformly in a batch furnace and the rate at which they can be ramped up and down in temperature. The final aspect comes from the technological strengths of RTP, which allow yield to rise when RTP is deployed in manufacturing [6, 21].

3. The evolution of RTP technology

Advances in wafer temperature control lie at the heart of improvements of RTP technology. The problem of temperature control can be split into three separate components. These are uniformity within the wafer, repeatability on different wafers of the same kind, and repeatability among wafers of arbitrary types, which may have radically different optical properties. This paper will demonstrate significant advances in RTP temperature control through a discussion of the principles of RTP system design, which will be illustrated using results obtained using the AG Associates 8800 Heatpulse, Starfire and Starfire 300 systems. The Heatpulse 8800 is a 'traditional' RTP system employing a crossed-linear-lamp array and a rectangular process chamber, while the Starfire and Starfire 300 tools represent an evolution in RTP technology which includes the use of wafer rotation, axisymmetric lamp arrays and multiple-point dynamic temperature control.

In the steady-state part of a heating cycle, wafer temperature uniformity depends on how the balance between the radiant power delivered and the heat loss varies with position on the wafer. Variations in the power delivered arise from the configuration of the heating lamps and the geometry of the reaction chamber [22]. Patterns in the coatings on the wafer, which may reflect varying amounts of the incident radiation, can introduce spatial variations in the power absorbed [23]. Heat loss also varies with position, as a result of the chamber geometry and the effects of patterns on the wafer. For example, the wafer edges tend to lose more heat than the center, partly because of the larger surface area at the edge and the fact that the edges may 'view' nonreflective features on the chamber walls and partly because the convective heat loss caused by process gas flowing over the wafer is largest at the edge [8]. Thermal conduction within the wafer modifies the impact of the imbalances between incident and radiated power fluxes on the temperature nonuniformity. Silicon is a good conductor of heat, and as a result, lateral thermal conduction within the wafer smoothes out the temperature profile, reducing temperature nonuniformity. The shorter the length scale on the wafer is, the stronger the effect of this smoothing is. This makes the temperature range within a length scale of ~1 mm unlikely to exceed ~1°C, but means that keeping the same temperature uniformity over a 10 cm length scale is much more difficult [23]. In this case there is effectively no smoothing from thermal conduction, and the extremes of the region behave as if they were different wafers, whose temperatures are defined by the local values for the power absorbed and emitted. The magnitude of the problem can be assessed by using a simple model for the power balance,

$$\eta P = H_{eff}\sigma T^4, \tag{1}$$

where η is the local power coupling constant, P is the local incident power density, σ is the Stefan–Boltzmann constant, T is the local wafer temperature and H_{eff} is a parameter describing the efficiency with which power is lost from the wafer by thermal radiation. The power may be incident from one side or both sides of the wafer and H_{eff} includes the possibility of loss from both sides of the wafer, as well as the effect of rereflection of the emitted radiation back onto the wafer. For a wafer in a black cavity, H_{eff} equals $2\varepsilon_{waf}$, where ε_{waf} is the average total hemispherical emissivity of the two sides of the wafer, since both surfaces can radiate freely without any rereflection effects. Eq. (1) can be used to determine the sensitivity of wafer temperature to fluctuations in η, H_{eff} and P. For fluctuations $\Delta\eta$, ΔH_{eff} and ΔP, the corresponding temperature fluctuation ΔT is given by

$$\frac{\Delta T}{T} = \frac{1}{4}\left(\frac{\Delta P}{P} + \frac{\Delta\eta}{\eta} - \frac{\Delta H_{eff}}{H_{eff}}\right). \tag{2}$$

This equation shows that variations in the P, η and H_{eff} have an equal impact on the temperature change. A 1% change in the incident power can be expected to produce a 0.25% change in the absolute temperature of the wafer. At 1100°C, the wafer temperature would change by $0.0025 \times (1100 + 273) = 3.4$°C. This example illustrates the formidable technical challenge presented by the need to control the 3σ wafer temperature uniformity to ~1°C.

One of the main practical consequences is the need for a carefully designed lamp heating array. The problem of designing a suitable heater is not the same as that of designing a very uniform illuminator. The system requires controllability as well as uniformity, where controllability refers to the ability to locally affect the temperature in one region of a wafer relative to another one, for example to counteract nonuniformity in the power loss distribution [22]. A heater with a multizone structure allowing for fine tuning of the wafer temperature profile can provide uniformity despite the differing power delivery requirements for various RTP recipes, different wafer types and the variations in heat transfer conditions caused by the use of various gases and flow rates. It also allows dynamic uniformity optimization during temperature ramps. Traditional RTP systems combine rectangular reaction chambers with linear lamp arrays, with different groups of lamps gathered together as zones, whose relative power settings can be adjusted to provide wafer temperature uniformity. This kind of approach has proved very successful for providing 3σ temperature uniformity ~ ± 5°C, which is suitable for 0.25 μm device fabrication, and would even meet the requirements for some 0.18 μm processes [24]. However, for the most demanding future RTP applications, further improvements in temperature uniformity are needed, and these can be obtained by developments such as the combination of axisymmetric lamp arrays and wafer rotation.

Wafer rotation smoothes out many of the local imbalances of power delivery and power loss which cause temperature nonuniformity. A simple example illustrates the power of wafer rotation in reducing nonuniformity. Fig. 4 shows a wafer in a chamber with some arbitrary lamp system, where one section of the chamber wall has a lower reflectivity than the rest, for example because of the presence of a wafer loading port. This 'dark' feature increases the value of H_{eff} for part of the wafer near it and locally reduces the temperature. As an example, consider a 200 mm wafer which has a 10 cm long arc at its edge facing the dark section of the wall, where H_{eff} is 5% larger than it is for rest of the wafer. Eq. (2) implies that this causes a

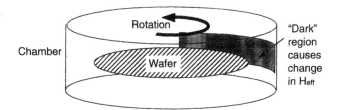

Fig. 4. An example of a disturbance in heat transfer conditions which could result in an azimuthal temperature nonuniformity on the wafer. A region of the chamber wall with a lower reflectivity locally increases the heat loss from the wafer. Rotation greatly reduces the thermal impact of azimuthal nonuniformities.

1.25% decrease in the temperature relative to that on the rest of the wafer, which means that when the rest of the wafer is at 1100°C, this segment is 17°C cooler, a serious nonuniformity. For a rotating wafer the problem can be analyzed by considering the behavior at a fixed position on the wafer and making H_{eff} vary with time as the wafer rotates. For simplicity, the model ignores the details of how the dark region comes into 'view' and assume that it appears and disappears suddenly once per revolution. H_{eff} is assumed to be unity when the point on the wafer is not facing the dark wall and to be 1.05 while it is in the 10 cm arc facing the dark zone. This period lasts for $30/(\pi R)$ s, where R is the rotation rate in revolutions per minute (rpm). A model is constructed using the equation

$$\rho c(T)D\frac{\mathrm{d}T}{\mathrm{d}t} = \eta P - H_{\text{eff}}(t)\sigma T^4, \qquad (3)$$

where ρ is the density of silicon, $c(T)$ is its temperature-dependent specific heat capacity, D is the wafer thickness, t is time and $H_{\text{eff}}(t)$ is the time-varying function describing the heat loss efficiency. Fig. 5 shows predictions for the temperature at a point on the wafer which is initially in the 'light' area of the chamber when rotation starts for various rotation rates. When the point moves into the dark zone its temperature

drops rapidly because of the increased radiation loss. Over the first few rotations there is a gradual decrease in the average temperature at the edge of the wafer because the increased radiation loss caused by the dark zone is spread out around the whole of the outer radius of the wafer. For all the rotation rates, the mean temperature becomes 1097°C, but the amplitude of the oscillation in temperature depends on the rotation rate, as shown in Fig. 6. The amplitude reduces from 17°C at the very low rates where the temperature reaches a steady-state condition in both the dark and the light zones, to almost zero at 200 rpm. Most of the benefit of rotation is achieved at rotation rates below ~30 rpm, above which there are diminishing returns in increasing the rate. The main benefit is that the rotation reduces a 17°C temperature nonuniformity in a particular region of the wafer to a 3°C drop at the edge, which is an axisymmetric feature which can be tuned out by adjusting the power distributed to the lamp zones. Effects of azimuthal nonuniformities in lamp illumination profiles and other heat transfer effects are also attenuated and turned into axisymmetric patterns.

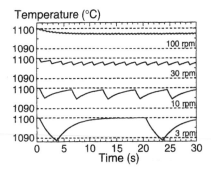

Fig. 5. The temperature observed at a fixed point on the wafer during rotation in the situation shown in Fig. 4. The point is initially in the 'light' zone of the chamber, and moves into the 'dark' region 0.5 s after rotation starts. The graphs show the behavior for various rotation rates.

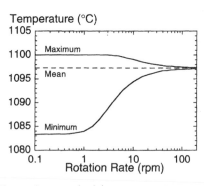

Fig. 6. The maximum and minimum temperatures experienced at a point on a wafer rotating in a chamber as shown in Fig. 4. At very low rotation rates, the point reaches a steady-state in both 'light' and 'dark' zones. As the rotation rate rises and the amplitude of the temperature oscillation decreases, the temperature stays close to the mean value of ~1097°C.

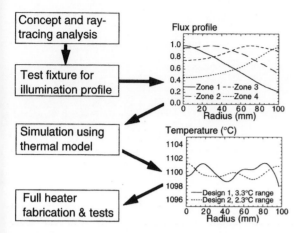

Fig. 7. The methodology used to design the Starfire heater. The combination of simulation techniques and test fixtures to evaluate the predicted effects of illumination profiles on wafer temperature uniformity and power requirements allows several design alternatives to be evaluated before the fabrication of full heaters for thermal and process testing.

The choice of the configuration of the lamp heating array is one of the most important aspects of RTP system design. In a linear-lamp RTP system the illumination profile is swept out over a very large area of the wafer as the wafer rotates under a lamp. While it is still possible to envisage designing a uniform power delivery system based on this configuration, it may prove difficult to provide the controllability needed to tune out radial temperature nonuniformities, because the great width of the averaged illumination profile from any lamp makes it difficult to correct for temperature gradients on length scales of a few cm. The design of an axisymmetric lamp array also poses many challenges, to ensure that the system can deliver the required uniformity and controllability, while still being efficient in power delivery to allow the system to heat wafers at high-ramp rates. The methodology used to design the Starfire system is schematically illustrated in Fig. 7. The optical design was configured with the help of ray-tracing models to predict the lamp zone illumination profiles and power transfer efficiency. The predictions were validated by using test fixtures which measure the optical power profile delivered to the plane of the wafer. Once these power profiles had been established, finite-element heat-transfer models were used to evaluate the consequences of given illumination profiles on wafer temperature uniformity and lamp power requirements. The rapid comparison of the temperature profiles predicted for many different potential designs allowed optimization of the lamp heating array before a real heater assembly was constructed. When the simulations indicated that the desired heater performance could be attained, the final stage of the development involved the construction of full-scale test

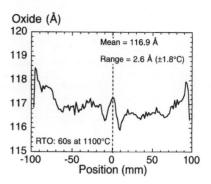

Fig. 8. Process uniformity obtained on a 200 mm wafer during rapid thermal oxidation at 1100°C for 60 s in the Starfire system. The result shown is from a 121-point diagonal scan of the oxide thickness, which implies a peak-to-peak temperature variation of ±1.8°C.

heaters, running process wafers and optimizing the lamp zone settings to obtain good process uniformity. Fig. 8 illustrates the 200 mm system capability by showing a cross-sectional scan of the oxide thickness grown on a wafer during a rapid thermal oxidation (RTO) experiment conducted at 1100°C for 60 s, with ramp rate of 75°C/s. The oxide thickness was measured by ellipsometry. The total range of thicknesses is 2.6 Å and the sensitivity of this process is ~0.72 Å/°C, which suggests a peak-to-peak temperature range of ±1.8°C. Fig. 9 illustrates the capability of the Starfire 300 system through a histogram of oxide thickness measurements from a series of six 300 mm wafers which were processed at 1100°C for 50 s in oxygen with a ramp rate of 75°C/s. The oxide thicknesses were measured at 121 points using a 3 mm edge exclusion. The 3σ-value for this distribution of oxide thicknesses is 1.95 Å, which is equivalent to a

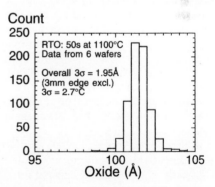

Fig. 9. The distribution of oxide thicknesses from a run of six 300 mm wafers processed in the Starfire 300 system. The results reflect the uniformity and repeatability during rapid thermal oxidation at 1100°C for 50 s and were obtained from 121-point oxide thickness contour maps measured with a 3 mm edge exclusion. The results demonstrate a 3σ distribution of 2.7°C.

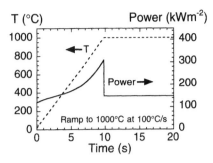

Fig. 10. The predicted power requirement for a ramp to 1000°C at 100°C/s. The peak power demand occurs just before the wafer reaches the steady state, when the power drops abruptly to the level needed to keep the wafer at 1000°C.

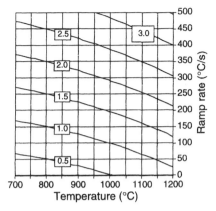

Fig. 11. Contour map of peak power requirements for linear ramps to various recipe temperatures. The results have been normalized to the peak power required when heating a wafer to 1000°C at 100°C/s as shown in Fig. 10.

combined uniformity and repeatability 3σ-value of 2.7°C, assuming a process sensitivity of 0.72 Å/°C.

One of the interesting challenges facing RTP equipment designers arises from the increasing need to provide the very high temperature ramp rates which are needed to minimize the effects of dopant diffusion during annealing of ion-implantation damage [25]. Eq. (3) can be used to predict the power required to heat the wafer to any given temperature at a fixed ramp rate by putting it in the form,

$$P = \frac{1}{\eta}\left(\rho c(T) D \frac{dT}{dt} + H_{eff}\sigma T^4\right). \quad (4)$$

Fig. 10 shows the predicted power profile required to heat a wafer to 1000°C at a constant ramp rate of 100°C/s. The prediction is for a 725 μm thick wafer with H_{eff} equal to unity, although in reality it depends on the configuration of the chamber and the nature of the wafer. η is also taken to be unity and the calculation takes no account of the effect of the lamp or chamber geometry on the efficiency of power delivery to the wafer. The peak power requirement occurs just before the wafer reaches the steady-state temperature in the recipe, where the power is about twice that needed to keep the wafer at 1000°C. Fig. 11 shows a contour map of the dependence of the peak power requirements on the steady state temperature and ramp rate. The map shows the ratio of peak power demand to that needed to heat to 1000°C at a rate of 100°C/s. In order to heat a wafer to 1000°C at 300°C/s one needs nearly twice the power required for 100°C/s, and for 500°C/s one needs more than three times as much power. The calculations show that the heater system must simultaneously achieve very high power delivery efficiency as well as good wafer temperature uniformity. Fig. 12 shows the results from a fast-heating experiment in the Starfire 300 system. The temperature was monitored using a pyrometer at the wafer

center. The graph includes the derivative of the pyrometer reading, which shows that the ramp rate was approximately constant at ~275°C/s as the wafer temperature rises from 600°C to ~970°C, at which point the lamps were switched off. The dynamic response of a wafer subjected to a constant power can be calculated using Eq. (3). In Fig. 12 the temperature rises approximately linearly with time because the power density delivered to the wafer, ηP, is fixed at a level which is very much greater than the heat loss term in Eq. (3), so dT/dt remains nearly constant.

The repeatability of the system is largely determined by the effectiveness of the closed-loop control on wafer temperature. In earlier generations of RTP equipment, closed-loop control was typically applied at only one point on the wafer, where a pyrometer would monitor the temperature and provide a feedback signal for control. The ratio of the power delivered to any lamp zone was kept fixed in any block of a processing recipe and the zones were 'slaved' together to respond to one feedback signal. Since the lateral thermal coupling only

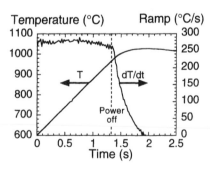

Fig. 12. Fast ramp in the Starfire 300 system. The results show the pyrometer reading and the ramp rate calculated from the temperature profile (dT/dt). The peak ramp rate is approximately 280°C/s.

extends a few cm at most from the control point, the rest of the wafer is not directly under closed-loop control. This requires the power delivery across the wafer to be inherently very stable with respect to that delivered at the control point. RTP systems are capable of giving very repeatable operation, even with this limitation, largely as a result of the use of high stability lamp power supplies, repeatable facilities and careful operation procedures. However, Eq. (1) dictates that for 0.5°C repeatability at 1100°C, the permissible variations in the power delivered, the power coupling and heat loss are less than 0.04%. This very tight specification indicates that automatic feedback control on all of the wafer temperature distribution is necessary to maintain long-term repeatable behavior. The problem can be solved by using several sensors to observe the wafer temperature at different radii and provide feedback to a multiple-input, multiple-output (MIMO) control algorithm which sets the power in each lamp zone so that all the temperature sensors remain on the required temperature trajectory. MIMO control is a powerful weapon in improving RTP repeatability, since automatic uniformity tuning compensates for subtle changes in heat-transfer conditions caused by the use of different recipes, the aging of system components and for variations in wafer properties. It also maintains temperature uniformity during the whole heating cycle, including the ramp stages, ensuring better process uniformity and minimizing the risk of slip on large diameter wafers. This simplifies the use of RTP because operators do not need to be skilled in adjusting lamp zone settings to maximize uniformity. Fig. 13 illustrates pyrometer readings from four pyrometers which provide feedback to a MIMO control algorithm during RTO performed at 1100°C with a ramp rate of 30°C/s. The temperature range between the pyrometer readings in the steady-state part of this cycle was ±2.5°C. During the ramp-up the spread was

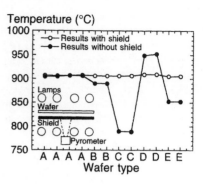

Fig. 14. Emissivity-independence obtained using a radiation shield approach in an AG Associates 8800 system [13]. The results show the temperature reported by a thermocouple on a selection of wafers with different backside coatings when the system was run under pyrometer control. When there is no shield, the emissivity variations cause a 160°C spread in temperature, but when the shield is present this reduces to 4°C.

±8°C at 800°C and ±4°C at 1050°C. The 49-point oxide thickness map for this wafer, which was measured using a 3 mm edge exclusion, indicated a 3σ uniformity of 1.7 Å on a mean oxide thickness of 83 Å (one point was excluded as a particle). The sensitivity for this process was previously established as ~0.58 Å/ °C, and the resulting 3σ-temperature uniformity is deduced to be 2.9°C.

Each temperature sensor channel must exhibit high temperature resolution and excellent stability. For the Starfire system a special new type of pyrometer was created, which samples radiation from several positions on the wafer via a network of fibre-optic cables, and provides temperature readings for each channel 50 times per second, to allow for excellent dynamic control at high ramp-rates. The pyrometer provides temperature readings from temperatures as low as 200°C, so that the full spectrum of RTP processes can be covered under closed-loop control.

The final element in the temperature control problem arises from the question of how to measure the temperatures of wafers with radically different optical properties. One of the stumbling-blocks of early RTP systems lay in their inability to process wafers which had different backside coatings without recalibration in order to correct for the impact of the coatings on the pyrometer monitoring the temperature of the wafer. Several approaches have been developed over the last few years which have greatly reduced the scale of this problem [11]. These include the use of a ceramic shield technology, which virtually eliminates the effect of wafer backside coatings on temperature control in the Heatpulse 8800 system, as shown in Fig. 14 [13]. Other promising developments in RTP temperature control include the use of ripple pyrometry, where the modulation of the lamp intensity is used to provide an in

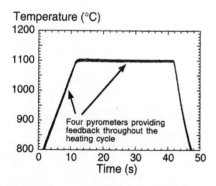

Fig. 13. Multiple-input, multiple-output (MIMO) control of a rapid thermal oxidation process at 1100°C. In the steady-state period, the pyrometers indicate a temperature range of ±2.5°C. For the case shown oxide thickness measurements suggest a temperature distribution with a 3σ of 2.9°C.

situ measurement of the reflectivity of the wafer, which can be used to correct for the emissivity of the wafer [14]. Another technique described recently involves the use of two sensors which view the same radius of a rotating wafer at different azimuthal locations [15]. The geometry of the two sensors is different, so that the effective emissivity of the wafer is different for the two sensors, allowing a manipulation of the two signals to produce an estimate of the wafer emissivity which can then be used to correct the temperature readings. Other developments in pyrometry, including a sophisticated implementation of the use of an auxiliary reflector to create a virtual black body cavity, have been incorporated into the Starfire system to provide emissivity-independent pyrometry.

These advances in temperature control demonstrate that RTP provides a unique capability for agile process development as well as robust manufacturing of advanced devices. The growth of RTP process applications will also arise from the remarkable flexibility of RTP systems for gas ambient control. One good example of this is in the rapid emergence of RTP as a tool for processing wafers in a steam ambient [26]. It has been shown that RTP steam oxidation can result in improved reliability in thin gate oxides, as well as providing a method for rapid growth of thicker oxide films for a number of process steps in advanced MOS technology. RTP has also been shown to provide new approaches for introducing nitrogen into thin oxides in order to improve reliability and decrease the penetration of boron atoms from polysilicon doped gate electrodes into the MOS channel [27]. The advantages of the controlled RTP ambient for processing metal films and silicides are well known, and the optimization of the gas ambient in RTP processing has been shown to lead to significant improvements in device yield [7, 21].

4. The future

The results and analysis presented in this paper illustrate the unique capabilities of RTP for meeting the semiconductor processing requirements of the next century and challenge the notion that batch furnace processing is in any way superior to RTP. In order to continue this success-story, RTP must clearly demonstrate its strengths in the manufacturing environment, including its inherent compatibility with the single-wafer processing environment which dominates the fabrication process, the fast cycle time which makes this technique a remarkable 'agile' approach to both development and manufacturing, and its unique process capabilities. Further technological developments in the basic heating and chamber technology can be expected to be combined with measures to simplify the

use of the equipment. Other long-anticipated developments include the combination of RTP with in situ wafer cleaning and with film deposition steps, for example in the formation of gate stacks [2]. As MOS-technology reaches its limits the control of the interfaces becomes critical, and significant improvements in device quality have been demonstrated by using these approaches [28].

Looking further into the future, it is already evident that RTP can provide solutions to many problems in emerging device technologies. In MOS technology, new materials and processes bring with them new thermal processing requirements, where the agility of RTP can be exploited to move rapidly from development to manufacturing. The emergence of quantum-effect based devices and single electronics as replacements for MOS technology offers new opportunities as a result of the remarkable control which will be needed to handle defects and atomic diffusion phenomena at nanometre length scales [29]. Applications can also be expected to expand in the field of microelectromechanical structures (MEMS) where control of film stress and mechanical properties has already created new RTP applications [30]. The development of new semiconductor materials for optoelectronics and high temperature electronics is also bringing with it a need for RTP [31]. RTP is also finding applications beyond semiconductor processing in other manufacturing processes, including magnetic materials development [32]. This spread of RTP concepts to create innovative manufacturing processes in diverse fields promises to be a very fruitful area of development in the next century.

5. List of symbols used in equations

η	lamp power coupling efficiency
P	lamp power density (Wm^{-2})
H_{eff}	power loss efficiency
σ	Stefan–Boltzmann constant ($Wm^{-2}K^{-4}$)
T	temperature (K)
t	time (s)
D	wafer thickness (m)
$c(T)$	specific heat capacity of silicon ($Jkg^{-1}K^{-1}$)
ρ	density of silicon (kgm^{-3})

Acknowledgements

The author would like to thank his colleagues at AG Associates for assistance in preparing this paper, and in particular JG Li, N Acharya and KS Balakrishnan, who provided experimental data.

References

[1] The National Technology Roadmap for Semiconductors, 1997 Edition, Semiconductor Industry Association, 1997.
[2] Wortman JJ, Ozturk MC Hauser JR. In: RTP'96, Fair RB, Green ML, Lojek B, Thakur RPS, editors. Round Rock: RTP'96 1996. p. 10.
[3] Lucovsky G, Hinds B. Mat Res Soc Symp Proc 1997;470:355.
[4] Chen PC, Hsu KYJ, Loferski JJ, Hwang HL. Mat Res Soc Symp Proc 1995;387:271.
[5] Pas MF, Pas SD. Mat Res Soc Symp Proc 1997;470:147.
[6] Thakur RPS, Schuegraf K, Fazan P, Rhodes H, Zahorik R. Mat Res Soc Symp Proc 1995;387:187.
[7] Miller M. In: RTP'93 Fair RB, Lojek B, editors. Scottsdale: RTP'93, 1993. p. 156.
[8] Roozeboom F. Mat Res Soc Symp Proc 1993;303:149.
[9] Leroy B, Plougonven C. J Electrochem Soc 1980;127:961.
[10] Timans PJ. In: RTP'96 Fair RB, Green ML, Lojek B, Thakur RPS, editors. Round Rock: RTP '96, 1996. p. 145.
[11] Timans PJ. Solid-State Technol 1997;40:63.
[12] Timans PJ. In: Roozeboom F, editor. Advances in rapid thermal and integrated processing. Dordrecht: Kluwer Academic Publishers, 1996. p. 35.
[13] Timans PJ, Morishige RN, Wasserman Y. Mat Res Soc Symp Proc 1997;470:57.
[14] Fiory AT. Mat Res Soc Symp Proc 1997;470:49.
[15] Peuse B, Yam M, Bahl S, Elia C. In: RTP'97 Fair RB, Green ML, Lojek B, Thakur RPS, editors. Round Rock: RTP '97, 1997. p. 358.
[16] Fan YH, Qiu T. Int J Heat Mass Transfer 1998;41:1549.
[17] Porter C, Laser A, Ratliff C. Mat Res Soc Symp Proc 1997;470:207.
[18] Van Schravendijk BJ, De Konig WL, Nuijen WC. J Appl Phys 1987;61:1620.
[19] Singh R, Sharangpani R, Cherukuri KC, Chen Y, Dawson DM, Poole KF, Rohatgi A, Narayan S, Thakur RPS. Mat Res Soc Symp Proc 1996;429:81.
[20] Hossain-Pas S, Pas MF. Mat Res Soc Symp Proc 1997;470:201.
[21] Weimer RA, Ratakonda D, Powell C, Nuttall M, Thakur RPS, Timans PJ, Shah N. In: Fair RB, Green ML, Lojek B, Thakur RPS, editors. Round Rock: RTP '97, 1997. p. 91.
[22] Spence PA, Winters WS, Kee RJ, Kermani A. In: RTP'94 Fair RB, Lojek B, editors. Round Rock: RTP '94, 1994. p. 139.
[23] Hebb JP, Jensen KF. J Electrochem Soc 1996;143:1142.
[24] Thakur RPS, Timans PJ, Tay SP. Solid-State Technol 1998;41:171.
[25] Shishiguchi S, Mineji A, Hayashi T, Saito, S. 1997 Symposium on VLSI Technology. Tokyo: Japan Society of Applied Physics, 1997. p. 89.
[26] Tanabe Y, Suzuki N, Natsuaki N. In: RTP'96 Fair RB, Green ML, Lojek B, Thakur RPS, editors. Round Rock: RTP '96, 1996. p. 5.
[27] Wristers D, Fulford J, Han LK, Chen T, Lin C, Chen K, Kwong, DL. In: RTP'95 Fair, RB, Lojek B, editors. Round Rock: RTP'95, 1995. p. 147.
[28] Sorsch TW, Baumann F, Boone, T, Green, ML, Rosamilia JM, Sapjeta J, Silverman PJ, Timp GL, Weir B. In: RTP '97, Fair RB, Green ML, Lojek B, Thakur RPS, editors. Round Rock: RTP'97. p. 150.
[29] Ahmed H. J Vac Sci Technol B 1997;15:2101.
[30] Garling SE, Koch DJ. In: RTP '96, Fair RB, Green ML, Lojek B, Thakur RPS, editors. Round Rock: RTP'96, 1996. p. 136.
[31] Sweatman D, Dimitrijev S, Li H-F, Tanner P, Harrison HB. Mat Res Soc Symp Proc 1997;470:413.
[32] Roozeboom F, Ruigrok JJM, Klaassens W, Kegel H, Falter M, Walk H. Mat Res Soc Symp Proc 1996;429:203.

PERGAMON

Materials Science in Semiconductor Processing 1 (1998) 181–186

MATERIALS
SCIENCE IN
SEMICONDUCTOR
PROCESSING

Modelling and Off-Line Optimization of a 300 mm Rapid Thermal Processing System

A. Tillmann*, S. Buschbaum, S. Frigge, U. Kreiser, D. Löffelmacher, T. Theilig, P. Schmid

STEAG AST Elektronik GmbH, Daimlerstrasse 10, 89160 Dornstadt, Germany

Abstract

The modelling of a new 300 mm rapid thermal processing (RTP) system is described. Conventional raytracing techniques are used to determine lamp intensity distributions on both 200 and 300 mm wafers. Simulation results are verified using the 'difference method' (difference between two process parameter distributions such as oxide thickness, where the absolute power of one single lamp is varied). Wafer rotation is incorporated in the model and its influence on the temperature distribution will be discussed. Off-line optimization of the temperature distribution is utilized using model-based control. Experimental results of implant annealing on both 200 and 300 mm are shown and critical parameters influencing the temperature uniformity are discussed. © 1999 Elsevier Science Ltd. All rights reserved.

1. Introduction

The requirements of the transition to 300 mm wafers, as well as the trend towards smaller device geometries (0.18 μm), lead to the development of a new rapid thermal processing (RTP) system AST3000, suitable for both 200 and 300 mm wafers. During the design phase of the tool three basic aspects were considered: (i) inherent uniformity of the heat flux onto the wafer, (ii) possibility of fast ramp up cycles (up to 350 K/s) and (iii) usability for both 200 and 300 mm wafers.

Equipment modelling was utilized during the design phase of the RTP chamber. Recommendations for chamber geometry as well as surface characteristics were given for the prototyping of the process chamber.

Since temperature uniformity remains one of the most critical parameters in RTP equipment (as well as aspects such as throughput, cost-of-ownership, etc.), we will present an approach to model temperature distribution on semiconductor wafers for both static and rotational processing. Experimental results will be compared with simulations and off-line optimization of the RTP chamber using model-based control is described. Experimental results of optimized recipes and lamp correction setups will be shown.

2. The AST3000 concept

The AST3000 is a new RTP tool developed for processing both 200 and 300 mm wafers. A new process chamber features dual-sided heating using two arrays of linear lamps, one above and one below the wafer. Each lamp has a maximum power of 3.1 kW, a total of 56 lamps is used. Dual side heating is necessary to achieve high ramp-up rates as well as to reduce pattern related temperature nonuniformity on the wafer. Also stress induced during ramp-up cycles can be reduced when heating the wafer from both sides. The chamber has water-cooled aluminum walls and a water-cooled stainless steel door. A leak-tight process chamber is obtained by separating the wafer from the atmosphere using two plane quartz plates and an inflatable seal

* Corresponding author. E-mail: a.tillmann@steag-ast.de

1369-8001/99/$ - see front matter © 1999 Elsevier Science Ltd. All rights reserved.
PII: S1369-8001(98)00029-8

(patent pending). The quartz plates are cooled using clean air.

The gas inlet is located opposite to the door. Laminar gas flow is achieved using a gas distribution system. The wafer is supported on a wafer tray consisting of an integrated rotation mechanism. Rotation is obtained by generating a process gas cushion below a centered, but moveable quartz plate. Nozzles outside the quartz plate projection can inject additional gas in order to accelerate or brake the rotation. All parts for the rotation mechanism are manufactured of quartz or sapphire and do not affect the temperature distribution on the wafer significantly. The parts for generating the process gas cushion are located outside the projection of a 300 mm wafer and therefore almost no shadowing on the wafer occurs.

During wafer rotation both rotation speed and exact angular position of the rotating quartz plate are monitored in real-time using optical sensors. This allows a separate controller to achieve and maintain any desired rotation speed and also to end the generation of the process gas cushion, when the rotation is finished at the end of the recipe. Rotation speeds up to 100 rpm can be used. However, due to the excellent basic uniformity (without lamp optimization) a rotation speed of about 20–40 rpm is high enough to guarantee best temperature uniformity.

Fig. 1 shows a cross-sectional drawing of the process chamber.

3. Model description

The model used in this study has been described earlier [1], therefore only a brief summary will be given. To model the heat transfer from the tungsten-halogen lamps to the wafer in the highly reflective enclosure, a raytracing algorithm was developed. The filament of the lamps is assumed to be cylindrical. From the surface of the cylinder a large number of rays is generated pointing into randomly selected

spatial directions. In this approach no spectral distribution of the intensity is used, only portions of the whole lamp intensity are assigned to the rays. After starting a ray, its path through the chamber is traced. If there is a reflection at the chamber walls, the intensity of the ray decreases by the portion absorbed by the walls. This portion is lost, since the chamber walls are water cooled. The quartz plates, which can have some influence on the distribution of light in the chamber is also neglected. Intersections of rays with filaments or quartz surroundings of other lamps are not taken into account. The good agreement between verification experiments and model predictions indicates that these assumptions are valid. The wafer is divided into ring segments. Each segment has the same area, the wafer flat or notch is neglected. 121 segments are always used. The whole wafer is assumed to be an infinite thin plate, having different reflectivities on the upper (polished) and lower (rough or polished) surface. It is assumed that these optical properties are constant across the whole surface. Any reflection at either wafer side is assumed to be specular. If a ray intersects any wafer segment the absorbed intensity at the wafer surface is added to the intensity assigned to this segment. The dependence of absorption on the incidence angle is neglected. Transmission of infrared light through silicon is negligible at higher temperatures. Intersections of rays with the edge guard ring are treated in the same way as intersections with the wafer, except that no intensity distribution on the ring is calculated. The algorithm uses two cut-off conditions together: a minimum of the remaining intensity and a maximum length of the ray. Typical values used during this study were 1% remaining intensity and 1 m maximum length.

The model must be configured accurately, therefore a realistic chamber layout has to be entered with correct position and size of the wafer, number and location of the lamps, chamber dimensions, optical properties of the reflecting surfaces, wafer and the edge guard ring. The raytracing algorithm can additionally be used to calculate the angle distribution of incident lamp light onto the wafer.

The intensity distribution on the wafer from each lamp is separately calculated and stored. Adding the modelled wafer radiation and silicon guard ring radiation the total absorbed intensity distribution is calculated. Based on this intensity distribution the temperature distribution can be calculated by numerically solving the heat conduction equation.

Wafer rotation is incorporated into the model with a simple technique: it is assumed that the temperature along one specific radius on the wafer is constant. Therefore an averaging of the temperature (and resulting process parameter) of radial zones allows for simulation of wafer rotation. Here it is necessary that the

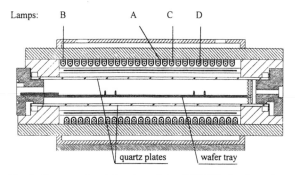

Lamps: B   A   C   D

quartz plates   wafer tray

Fig. 1. Cross sectional drawing of the AST3000 chamber. A–D are denoted lamps which are used in Section 3.

rotation speed is high enough so that any changes in temperature distribution have a longer duration than one rotation. In case of circular site patterns this averaging can be carried out rather easily. In case of cartesian site patterns a special mathematical algorithm should be used for averaging as it is incorporated e.g. in WAFERMAP software [2].

4. Comparison of simulated and experimental data

To verify the model, so-called 'difference maps' have been calculated. A difference map can be obtained by subtracting two different sets of data generated on a four point probe or ellipsometer. The first set of data represents any standard process. The second represents a similar process, where one parameter is changed, e.g. one single lamp is switched off. It is important to keep all other processing parameters constant. The difference map will visualize the influence of this single parameter on the wafer. In the following sections we will distinguish between static and rotational processing.

4.1. Static processing

For the following experiments an anneal process at 985°C for 20 s has been chosen for an As implant performed at an energy of 20 keV and a dose of 1E16/cm^2 on 200 mm wafers. 121 measurement sites were used as standard set-up to measure sheet resistance (R_s) distribution with an edge exclusion of 3 mm. For the comparison of simulation and experimental results special lamps have been selected, which represent most probably the worst cases regarding influence on temperature uniformity. These lamps are marked in Fig. 1 as lamp A (center of wafer), lamp B (outside wafer projection), lamp C (between wafer center and edge) and lamp D (at wafer edge). As mentioned above, difference maps are used to visualize the influence of single lamps on the wafer. These difference maps are generated from both experimental sheet resistance distributions and simulated ones. Fig. 2 shows the results. All distributions are displayed with a 0.2 Ω/sq contour interval. Generally there is a good agreement between model and experiment.

4.2. Processing using wafer rotation

The same anneal process as for static processing has been used and also the same lamps have been selected for the difference maps. For these experiments an older version of the rotation mechanism has been used, which allowed for up to 10 rpm. Therefore, the resultant maps did not show perfect rotational symmetry. Because of the missing symmetry the measured R_s distributions were averaged along radial zones. This gives

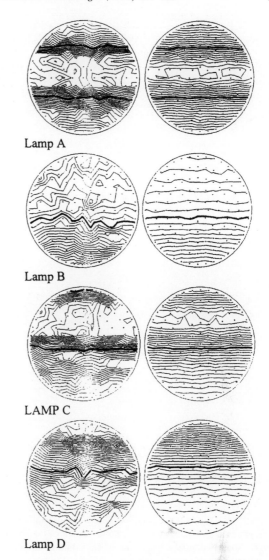

Lamp A

Lamp B

LAMP C

Lamp D

Fig. 2. Difference maps to visualize influence of single lamps on the wafer. Left: experimental results. Right: simulated results for lamps A–D. Contour interval is 0.2 Ω/sq.

a clearer picture of the results. Fig. 3 shows the results. All distributions are again displayed with a 0.2 Ω/sq contour interval. Again, a sufficient agreement between model and experiment can be found.

The results prove, that the presented model is valid. It should be emphasized that a very simple mathematical model was developed which neglects all quartzware in the processing chamber, as well as any convective effects from the process gas. Only heat radiation is incorporated into the model.

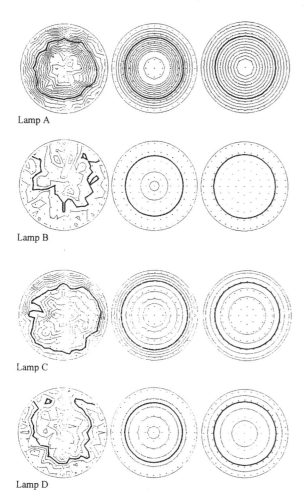

Lamp A

Lamp B

Lamp C

Lamp D

Fig. 3. Difference maps to visualize influence of single lamps on the wafer during wafer rotation. Left: experimental results. Center: radially averaged maps of the experimental results. Right: simulated results for lamps A–D. Contour interval is 0.2 Ω/sq.

5. Optimization of temperature uniformity

Optimization of temperature uniformity can be done by either manual or automatic tuning of lamp correction tables in the system described. It has been shown earlier [2], that model-based equipment control can lead to excellent temperature uniformity results. The optimization algorithm OPUS can easily be configured for any lamp configuration and was tested on the AST3000.

To demonstrate the ability of the AST3000 to provide exceptional temperature uniformity during ramp-up and steady state conditions, a set of experiments has been performed. As mentioned above, one basic aspect during the design phase of the process chamber was to design a chamber with an excellent inherent temperature uniformity. Because this is also the start-

ing point for each OPUS optimization, the uniformity has to be as good as possible, because the level of possible compensation using a lamp correction table (LCT) generated by OPUS is also limited. This is due to the soft gradient of the influence of a single lamp onto the temperature distribution on the wafer (Fig. 3). Only by this soft gradient a real good inherent uniformity can be reached without a visible influence of each single lamp. This means no reflectors or lightpipes are allowed [3]. Using wafer rotation even a bad uniformity can be compensated by averaging the nonuniformities [4], however this procedure may cause a stress on the wafer. This method is also limited to processes with a certain soak time, it will not be the appropriate way for flash anneal processes which are more critical.

The following experiments were made using As, 1E16, 20 keV implanted 200 and 300 mm wafers. The anneal conditions were 1000°C, 30 s. This resulted in an R_s of approximately 75 Ω/sq at a sensitivity of around 0.76 Ω/sq/K. In the following all standard deviations (S.S.) refer to the R_s. Fig. 4 shows the excellent basic uniformity of the process chamber of the AST3000. On 200 mm wafers a temperature range (maximum–minimum) of 4.8 K at a S.D. of 1.1% could be reached (test diameter 194 mm). The wafer was annealed without applying any lamp correction. By simply rotating the wafer with 40 rpm the temperature range across the wafer could be reduced to 3.7 K with again 1.1% in S.D. This indicates obviously a non-Gaussian distribution of the R_s values on a rotated wafer. A follow-up OPUS optimization resulted in 2.4 K at 0.77% of S.D.

Another experiment was performed using so-called 'flash processes'. Again only one LCT was used during the entire process. The ramp-up rate was 200 K/s, the cool down rate around 80 K/s, which resulted in an R_s of 129.4 Ω/sq. The temperature range across the wafer in that case was 4 K, assuming the same sensitivity as

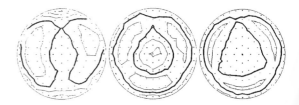

Fig. 4. 1E16, As, 20 keV implanted 200 mm wafer after annealing. The process conditions were 1000°C, 30 s steady state duration. Contour interval is 1%, 194 mm test diameter. Left: the inherent uniformity of the system without wafer rotation and without applying any lamp correction. This results in a temperature range of 4.8 K at 1.1% S.D. Center: using wafer rotation the uniformity was improved to 3.7 K range with 1.1% S.D. Right: application of OPUS-optimization leads to 2.4 K range at 0.77% S.D.

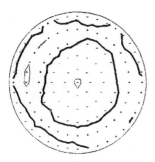

Fig. 5. R_s distribution of an 1E16, As, 20 keV 200 mm wafer after a flash anneal with a ramp up rate of 200°C/s up to 1000 K/s. Contour interval is 1%, 194 mm test diameter. The temperature range across the wafer is 4 K. S.D. is 0.53%.

Table 1
Resultant standard deviation and range of R_s after implant anneal during static and rotational processing, when various lamps are switched off

Condition	Static processing		Rotational processing	
	S.D. (%)	range (Ω/sq)	S.D. (%)	range (Ω/sq)
Starting LCT	2.75	8.9	0.81	2.6
Lamp A off	2.72	9.3	1.57	4.3
Lamp B off	4.02	10.0	0.65	2.2
Lamp C off	2.02	6.7	1.0	3.0
Lamp D off	2.43	8.2	0.66	2.7

above. The S.D. was 0.53%. The result is shown in Fig. 5. The low temperature range across the wafer, the concentrical shape of the R_s distribution and the fact of needing only one LCT for the whole process indicates a very low dynamic temperature gradient. The steady state duration was <1 s.

On 300 mm wafers the inherent uniformity is not as good as on 200 mm wafers because of the larger wafer size. An implant anneal results in 4.3% S.D. and 20.8 K across the wafer, calculated from the R_s distribution after an implant anneal of 1000°C, 30 s applying no lamp correction. Using wafer rotation this value is reduced to 10.1 K, 3.6% S.D. By further OPUS optimization a 2.5 K range across the wafer and 0.76% S.D. could be realized. All measurements were performed with an edge exclusion of 5.5 mm following the specifications of the implanters vendor. Fig. 6 shows the results. There is still room for further improvement of the inherent temperature uniformity and future development will focus on optimizing the wafer position in the chamber and to reduce the center to edge gradient.

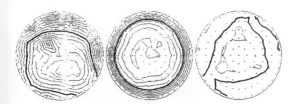

Fig. 6. 1E16, As, 20 keV implanted 300 mm wafers used to monitor the temperature distribution of the AST 3000 system. The process conditions were 1000°C, 30 s steady state duration, the measurements were performed with a test diameter of 289 mm at 121 sites. The contour interval is 1%, Left: the inherent uniformity of the system without wafer rotation and applying any lamp correction results in a temperature range of 20.8 K with 4.3% S.D. of the mean R_s. Center: using wafer rotation a range of 10.1 K, at a S.D. of 3.6% was achieved. Right: after optimization a range of 2.5 K at a S.D. of 0.76% across the wafer has been demonstrated.

Due to the very small influence on the temperature uniformity from a majority of lamps the whole system can guarantee very stable and reproducable results. This was verified by a set of experiments, where again some lamps were switched off. This simulates a worst case condition during processing: a lamp failure. The same anneal process as described above at 985°C for 20 s has been chosen for an As implant done at an energy of 20 keV and a dose of 1E16/cm^2 on 200 mm wafers. 121 measurement sites were used as standard set-up to measure sheet resistance distribution with an edge exclusion of 3 mm. The same lamps as described above have been chosen for these experiments. The resultant standard deviation of the R_s distribution and the range (maximum–minimum) are shown in Table 1.

It is not surprising that switching off of these lamps does not disturb the good temperature uniformity (except for the center lamp) during wafer rotation. In this case the influence of single lamps becomes much smaller than for static processing. This can be explained from the difference maps shown above and is due to the reduction of a real 2-D lamp intensity distribution down to a 1-D distribution. This guarantees very stable processing results with almost no drift in the temperature uniformity over a long period of time. It is expected from the majority of all other lamps, especially outside the wafer projection, that there is a similar small influence on the temperature uniformity during wafer rotation.

6. Conclusion

A new rapid thermal processing system has been presented, which allows for processing of both 200 and 300 mm wafers. The system uses wafer rotation during processing, which is achieved by a process gas cushion below a rotating quartz plate. Raytracing is used to calculate the intensity distributions of the lamps on the wafer. Difference maps have been used to verify simulation results. There is a good agreement between experiment and simulation and therefore the intensity distributions can be used in a mathematical model of the RTP chamber. Experimental results of off-line optimization of the chamber have been shown. Best uniformity results were a range of 2.4 K for 200 mm and 2.5 K for 300 mm wafer.

Acknowledgements

The authors wold like to acknowledge the following people for their support, discussions and help during preparation of this paper: Dr. Zsolt Nényei, Dr. Wilfried Lerch, Wayne Morrow, Cheryl Boots and Heinz Walk. The authors spent several hours of interesting and productive discussions with Dr. Kersch at Siemens and would like to thank him for helpful input.

Special thanks to STEAG AST's application laboratory staff for their help in performing experiments and measurements.

References

[1] Tillmann A. Model based temperature uniformity control during rapid thermal processing. In: Proceedings RTP'96.
[2] Application Note 6 'Simulation of wafer rotation'; Tomerdingen, Germany: Boin GmbH.
[3] Tillmann A, et al., Transient thermal behavior in a new RTP chamber. In: Proceedings TMS'98.
[4] Nagabushnam RV, Singh RK, Sharan S, Sandhu G. Mat Res Soc Symp Proc 1996;429:95.

PERGAMON

Materials Science in Semiconductor Processing 1 (1998) 187–193

MATERIALS
SCIENCE IN
SEMICONDUCTOR
PROCESSING

Perspectives on emissivity measurements and modeling in silicon

S. Abedrabbo [a],*, J.C. Hensel [a], A.T. Fiory [b], B. Sopori [c], W. Chen [a,c], N.M. Ravindra [a,1]

[a]*New Jersey Institute of Technology, Newark, NJ 07102, USA*
[b]*Bell Laboratories, Lucent Technologies, Murray Hill, NJ 07974, USA*
[c]*National Renewable Energy Laboratory, Golden, CO 80401, USA*

Abstract

A spectral emissometer operating in the wavelength range of 1–20 μm and temperature range of 30–900°C has been utilized to simultaneously measure the reflectance, transmittance and emittance of silicon. Interesting differences in the optical properties have been reported due to differences in surface morphology. Quantitative results of the effects of rough side incidence versus smooth side on the optical properties of the same silicon wafer are analyzed in this study. This analysis is based on a standard one-parameter, multiple-reflection model as extended by Vandenabeele and Maex to include effects of a roughened surface. Their modification essentially replaces the usual internal attenuation factor by an enhanced effective attenuation factor to take into account the effects of surface roughness. In the present study, it has been found that this very simple model gives a good account of the optical properties when radiation is incident on the smooth side of the wafer but fails for incidence for the roughside. © 1999 Elsevier Science Ltd. All rights reserved.

1. Introduction

For rapid thermal processing (RTP), pyrometers are the instruments of choice for in situ temperature measurements. Pyrometers measure the amount of radiation emitted from a wafer within a narrow wavelength window. The ratio of the wafer emitted radiation to that of a blackbody under the same conditions of temperature, wavelength, angle of incidence and direction of polarization is referred to as emissivity. Emissivity of silicon is a complicated func-tion of both temperature and wavelength [1]. It is also a function of surface roughness [2]. The temperature and wavelength dependent emissivity of silicon has been studied extensively in the literature. However, studies of optical properties as a function of surface roughness that covers a continuous range of wavelengths in the infrared is lacking. To the best of our knowledge only one paper [3] has tackled the subject with emphasis on the backside roughness effect on the emissivity of silicon in range of temperatures from 300–700°C at two specific pyrometric wavelengths of 1.7 and 3.4 μm.

In the present study radiative properties for silicon sample with single side polish have been investigated for a wide and continuous range of wavelengths from 1–20 μm. Interpretation for the measured properties

* Corresponding author.
[1] Tel.: + 1-201-596-3278; fax: + 1-201-642-4978; e-mail: ravindra@admin.njit.edu

1369-8001/99/$ - see front matter © 1999 Elsevier Science Ltd. All rights reserved.
PII: S1369-8001(98)00028-6

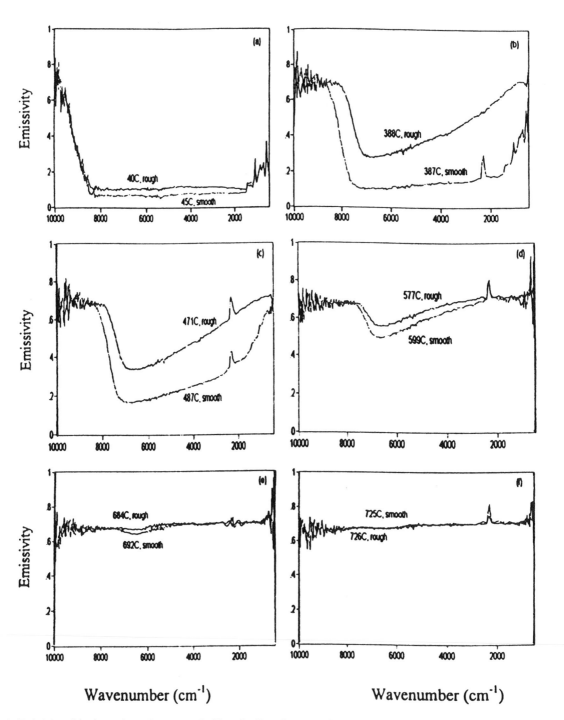

Fig. 1. Emissivity of back-rough vs. front smooth sides of *n*-Si as function of Wavenumber for specific temperatures: (a) 40, 45°C (b) 388, 387°C (c) 471, 487°C (d) 577, 599°C (e) 684, 692°C and (f) 726, 725°C, respectively.

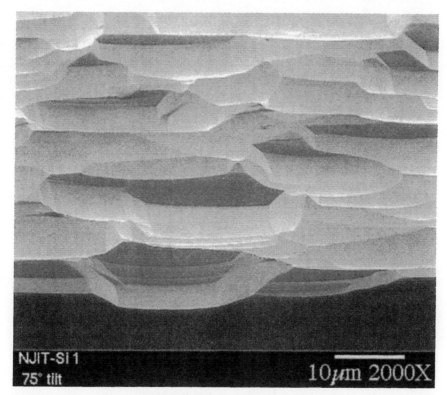

Fig. 2. SEM (2000X) picture of the rough side of n-Si.

have been sought with the help of the model proposed by Vandenabeele and Maex.

2. Experimental details

The spectral emissometer utilized in this study has been discussed earlier [4, 5]. It consists of an hemi-ellipsoidal mirror providing two foci, one for the exciting source in the form of a diffuse radiating near-blackbody source and the other for the sample under investigation. A microprocessor controlled motorized chopper facilitates simultaneous measurement of sample spectral properties such as emittance, reflectance and transmittance. A carefully adjusted set of five mirrors provides the optical path for measurement of the optical properties. The source of heating of the samples is provided by an oxy-acetylene/propane torch. The spectral emissometer utilizes the Helmholtz reciprocity principle [6] as has been explained in a related study [7].

3. Results and discussion

In this study the radiative properties of standard silicon wafers with an industrial grade rough back-

side of the order of ~8 μm peak to valley (~1 μm rms) [8] have been studied extensively as a function of temperature. The wafer under investigation is n-type crystalline silicon, 650 μm thick, of resistivity = 8 Ω·cm.

In Figs. 1(a–f) the emissivity as measured from the front polished side and the rough side of the silicon wafer as a function of wavelength at specific temperatures is shown. As can be seen in this figure, the emissivity of the rough side is greater than that of the polished side. This remains the case until the sample becomes opaque to subbandgap radiation at temperatures above 700°C. In the range of temperatures investigated, the greatest difference in emissivities is observed at 387°C. Even though changing the surface morphology does not affect the optical constants of materials, i.e. refractive index n and extinction coefficient k, it does however change the absorptance of the substrate as has been described in detail in a recent paper [9].

In Fig. 2 the representative SEM micrograph of the sample under consideration is presented. The micrograph depicts the extreme peak to valley roughness in this sample. However, it shows clearly a high degree of flatness over large areas. This could explain the slight differences between the rough side incidence and the smooth side incidence measured reflectance, as was observed for this sample [9]. As the light is incident on such a specular surface it does not suffer unexpected scattering and therefore the intensity of the reflected light detected from the rough side is essentially the same as that of the smooth side.

3.1. The Vandenabeele Maex (VM) model and its application

In their work Vandenabeeble and Maex [2, 3] have found for incidence on the rough side that the influence of roughness on the emissivity is different for par

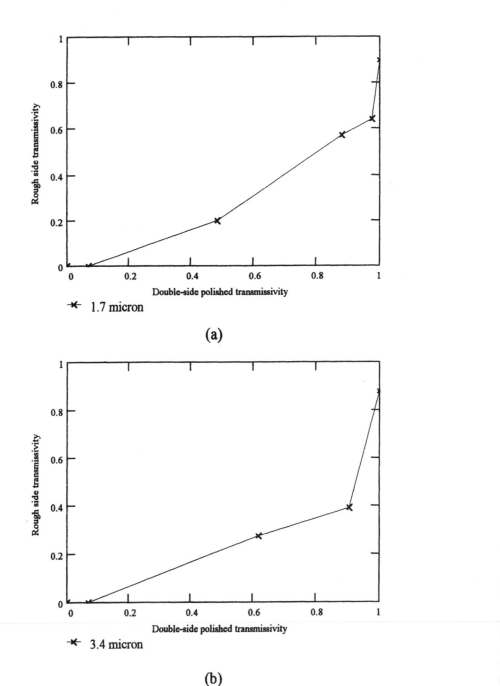

(a)

(b)

Fig. 3. Correlation between the true transmittance of a simulated double sided polished wafer and of a measured rough side of single side polished wafer for (a) 1.7 and (b) 3.4 μm.

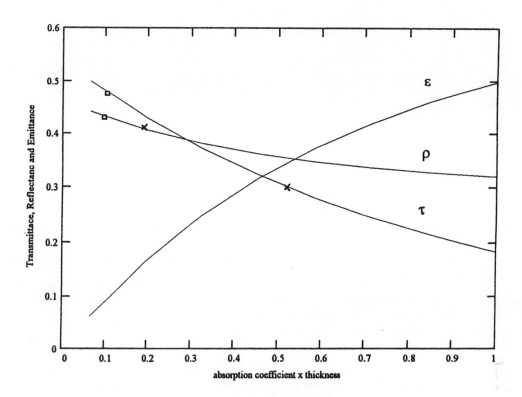

×   Rough side reflectance and transmittance at 6000cm-1
□   Smooth side reflectance and transmittance at 6000cm-1

Fig. 4. Demonstration of the one parameter model by comparing the experimental reflectance and transmittance at 6000 cm^{-1} of the rough incidence with that of the smooth incidence of n-Si sample, 0.065 cm thick, as function of the dimensionless parameter (absorption coefficient × thickness).

tially transparent and opaque wafers. For opaque wafers, only a slight dependence of the emissivity on the backside roughness is observed, except for an extremely rough backside. For partially transparent wafers, a strong dependence of the emissivity on the backside roughness is observed. They modeled this by defining an effective attenuation factor (rough side true transmissivity) $\tau_r(\lambda)^*$ [1, 3], which can be written as: $\tau_r(\lambda)^* = F\tau_r = Fe^{-\alpha t}$, where α is the absorption coefficient and t the effective sample thickness, the factor F lies between 1 and 0 depending on backside roughness. They find that the factor F is constant and represents the ratio between the true transmissivity of the rough side τ_r^* of the single side polished sample to that of the double side polished sample τ_r. They have shown that this model works well at 1.7 and 3.4 μm. In Fig. 3(a and b), results are shown of the application of the VM model to our data. The true transmissivity of the double side polish (DSP) wafer as function of temperature in Figs. 3(a and b) has been estimated for our

sample using the MIT/SEMATECH Multi-RAD model for the specifications of the n-Si considered in this study. As can be seen in this figure, the rough side true transmissivity does not vary linearly with that of the DSP wafer. This result is the same for 1.7 and 3.5 μm. The ratio between the two true transmissivities, F, is an exponential function and therefore the plot is indeed indicative of this behavior. In the investigation by VM, a linear relation has been proposed between the two transmissivities by excluding the true transmissivity values exceeding 0.82. In their study VM have suggested that above this value the trapped light will eventually escape after many internal reflections. Further, Fig. 3 shows that the plot of τ_r^* versus τ_r does not pass through the origin for 1.7 μm. This result is expected because with increasing α (but temperatures below opacity) the simple relationship above does not hold because the much longer effective internal path of the light for the rough side will cause full internal attenuation.

Table 1
Results of one parameter model in matching the experimental data in 1.67 μm

	Rough	($\alpha \times$ thickness)	Smooth	($\alpha \times$ thickness)
Transmissivity	0.3	0.52	0.47	0.104
Reflectivity	0.41	0.195	0.43	0.0975

3.2. One-parameter model

As mentioned above, the standard, multiple-reflection model is a one-parameter model, the parameter being αt (or alternatively, $\tau_r = e^{-\alpha t}$). In the work of VM the application of this model has been limited to just one measured optical property, the emissivity ε:

$$\varepsilon = (1 - \rho)(1 - \tau_r)/(1 - \rho \tau_r) \qquad (1)$$

where ρ is the "true" reflectivity of the 1st surface. This quantity ρ can be determined from measurements of emissivity ε at temperatures above the onset of opacity ($\tau_r \ll 1$). With ρ so determined substitution of data for ε at temperatures below opacity yield empirical values for τ_r. This standard model was derived for conditions of normal incidence on a wafer with plane, parallel, smooth surfaces. VM have extended the model to the case of rough surfaces by essentially making the replacement $\tau_r \rightarrow \tau_r^*$. To test this approximation we examine separately two other optical properties derivable from this same standard, one-parameter model, namely reflectance R and transmittance T:

$$R = \rho\{1 + \tau_r^2(1 - \rho)^2/(1 - \tau_r \rho^2)\} \qquad (2)$$

$$T = \tau_r(1 - \rho)^2/(1 - \tau_r^2 \rho^2) \qquad (3)$$

(The three-emittance, reflectance and transmittance-are, of course, not independent, being related by Kirchoff's relation). We calculate R and T as a function of the one parameter αt, the results being shown in Fig. 4 for $\rho = 0.300$. (In a sample with rough surfaces we can regard t as an "effective" thickness > actual thickness, if α is the actual absorption coefficient of the material). Consider now data [9] for reflectance and transmittance measured at 388°C and 6000 cm^{-1} (1.67 μm) listed in Table 1 and plotted on the appropriate curves in Fig. 4.

The corresponding values of αt determined in Fig. 4 are given in Table 1. It is clear from Fig. 4 and from Table 1 that transmittance and reflectance for the given experimental conditions are consistent with a single value of the parameter αt for smooth side incidence whereas they are seriously inconsistent for rough side incidence. We can conclude that the one-parameter model is valid for the former case but not

the latter. One needs to confirm this at other wave lengths.

4. Conclusions

A spectral emissometer that utilizes the Helmhol reciprocity principle has been used to acquire the opt cal properties of single side polished silicon wafe Surface roughness has an important effect on the emi sivity of the wafer. The emissivity of the unpolishe side shows an enhancement over that measured wit the polished side in the extrinsic temperature regim The greatest difference in emissivities in front-polishe versus back-rough is observed in the 387°C data and only weakly dependent on wavelength in the mid-I range. In order to understand the influence of surfac roughness on the optical properties of silicon, the sta dard model used and extended by Vandenabeele ar Maex has been utilized extensively. Limitations of th one-parameter model have been presented.

Acknowledgements

This work was supported partially by grants fro the New Jersey Commission on Science ar Technology, SEMATECH contract No. 36022090 US Air Force Wright Laboratory and the DARP (TRP) Microelectronics Technology Office under co tract F33615-92-C-5817, DARPA (TRP) contra DAAH04-94-C-0041 and DARPA contract DAAH0 95-1-0056 awarded through US Army Research Offic Research Triangle Park, NC. The authors acknowled with thanks the support of Burleigh Instruments obtaining roughness measurements.

References

[1] Sato K. Jpn. J. Appl. Phys. 1967;6(3):339–47.
[2] Vandenabeele P, Maex K. SPIE Proc. 1990;1393:316–3(
[3] Vandenabeele P, Maex K. J. Appl. Phys. 1992;72:586 75.
[4] Ravindra NM, Abedrabbo S. Chen W, Tong Fl Nanda AK, Speranza T. IEEE Trans. Semicond. Man 1998;Feb:30–9.

[5] Ravindra NM, Tong FM, Schmidt W, Chen W, Abedrabbo S, Nanda A, Speranza T, Tello AM. ISSM'96, Tokyo, Japan. 1996;Oct.:101–4.

[6] Born M., Wolf E., Principles of optics, Chap. 8.3, 4th ed., Pergamon Press, 1970. p. 381.

[7] Ravindra N.M., Abedrabbo S., Gokce O.H., Tong F.M., Patel A., Rajasekhar V., Williamson G., Maszara W., IEEE Trans. Components Packaging Manuf. Technol., 1998; Sept.;21:441–9.

[8] Measurements of roughness performed at Burleigh Instruments, April, 1998.

[9] Abedrabbo S., Hensel J.C., Gokce O.H., Tong F.M., Sopori B., Fiory A.T., Ravindra N.M., Mat. Res. Soc. Proc., 1998;525:95–102.

PERGAMON

Materials Science in Semiconductor Processing 1 (1998) 195–200

MATERIALS
SCIENCE IN
SEMICONDUCTOR
PROCESSING

New methods of metrology data analysis during semiconductor processing and application to rapid thermal processing

Manuela Boin [a],*, Wilfried Lerch [b],[1]

[a]*Boin GmbH, Graf-Albrecht-Str. 24, 89160 Tomerdingen, Germany*
[b]*STEAG AST Elektronik GmbH, Daimlerstr.10, 89160 Dornstadt, Germany*

Abstract

New mathematical methods for data analysis are presented in this paper. These methods are applied to data derived by semiconductor metrology tools. Sheet resistance data of implant anneal processes on 200 mm wafers done in a rapid thermal processing system are analysed for implanter and RTP tool inhomogeneities.

The first method, a separation method, is used to visualise uniformity of semiconductor process equipment (e.g. rapid thermal processors). The second method, also a separation method, is used to analyse input or preprocess uniformity of process-influencing parameters on the wafer (e.g. ion implant systems). Furthermore a subtraction method is shown which provides some additional advantages.

Basis for the data processing is a standardised interface developed to import and analyse data files from different metrology equipment such as ellipsometers and four point probes. © 1999 Elsevier Science Ltd. All rights reserved.

1. Introduction

The demands on the resultant homogeneity of physical parameters such as sheet resistance (or sheet thickness) increase with increasing wafer size (150, 200, 300 mm) and decreasing structure dimensions (0.25, 0.18, 0.13 μm). To be able to increase the homogeneity of these parameters it is important to distinguish between different sources of inhomogeneity. Distinguishing between different sources of nonuniformity of physical parameters on wafers results in an increasing importance of data analysis and data processing.

The resultant homogeneity of physical parameters on a wafer after any process is influenced by the superposition of the uniformities of all processes the wafer already went through (denoted as preprocess uniformity hereafter) and of course the uniformity of the process tool itself (denoted as tool or process uniformity). To be able to improve the homogeneity of these physical parameters on the wafer, it is important to identify the uniformity of single influences. If these uniformities are determined an optimization of the influence with highest nonuniformity can start.

The presented separation methods enable the user to split the nonuniformity of measured parameters on the wafer into wafer preprocessing and tool influences. Especially if both preprocess and tool uniformities can not be measured directly (e.g. in a rapid thermal annealing process after ion implantation) it is difficult to improve the resultant uniformity. In this paper the distribution of sheet resistance of wafers after rapid thermal annealing (RTA) of ion implantation of these wafers is used to demonstrate the methods by separating both uniformities of RTP tool (tool) and the ion implanter (preprocess).

* Corresponding author. E-mail: manuela.boin@boin-gmbh.com.
[1] E-mail: w.lerch@steag-ast.de

1369-8001/99/$ - see front matter © 1999 Elsevier Science Ltd. All rights reserved.
PII: S1369-8001(98)00043-2

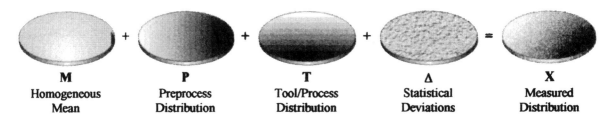

M	**P**	**T**	**Δ**	**X**
Homogeneous	Preprocess	Tool/Process	Statistical	Measured
Mean	Distribution	Distribution	Deviations	Distribution

Fig. 1. Example of the superposition of the contributing distributions to the overall measured parameter distribution.

2. Methods

It is assumed that a preprocess or input data set is changed by an arbitrary, but fixed process. The resultant output distribution is measured. These methods are not limited to applications in the semiconductor industry. They can be applied to any measurable set of parameters. Nevertheless, in this paper the methods are demonstrated using data measured on silicon wafers.

The measured parameter distribution, X, generally can be split into 4 main influences or parts: a homogeneous mean of the measured physical parameter, M, the deviation from this mean caused by all preceding processes ('preprocess uniformity'), P, the deviation from this mean caused by the used process tool ('tool uniformity'), T, and the sum of all statistical deviations from the average tool and preprocess uniformities, Δ. A linear superposition of preprocess and tool uniformities is assumed here.

$$X = M + P + T + \Delta \tag{1}$$

Fig. 1 visualises the superposition of the contributing distributions to the overall measured parameter distribution.

The basis for the new, very general analysis methods of metrology data is the availability of data sets processed and measured under the following conditions:

1. Same, repeatable preprocesses for all wafers (e.g. same wafer type and same implanter).
2. Same recipe used for the investigated tool during repeatable processing of all analysed wafers.
3. All wafers processed at different alignment angles α (see Fig. 2), which should be a fraction of $360°$ (e.g. 1/8, respectively 45°, 1/6 respectively 60°, or 1/4, respectively 90°), in one of the tools.
4. All wafers measured repeatably with same orientation of the flat or notch in the measurement tool, or the rotation of the measured parameter distribution afterwards using tools such as WAFERMAP [1].

For the usage of the methods only two wafers processed under these conditions are necessary, because the methods were developed based on the availability of two data sets. The availability of more data sets can be useful for cross checking and allows for averaging the results of different measurements to reduce statistical deviations.

In the following, the methods will be described based on two wafers processed at different alignment angles in the tool. Assuming two data sets measured as described above Eq. (1) leads to the following set of equations. Both data sets consists of the four contributing distributions M, P, T and Δ, but they might differ.

$$X_1 = M_1 + P_1 + T_1 + \Delta_1, \tag{2a}$$

$$X_2 = M_2 + P_2 + T_2 + \Delta_2. \tag{2b}$$

A detailed analysis of these single contributions shows that the mean values and the preprocess uniformity of both distributions are the same if we can assume repeatable wafer preprocessing and also repeatability of the processes of the investigated tool.

$$M_1 = M_2 = M \tag{3a}$$

$$P_1 = P_2 = P \tag{3b}$$

An analysis of the tool influence on these wafers identifies that it is basically the same if repeatability can be assumed. The only difference is that this distribution is rotated by an angle $-\alpha$ in the second data set. This angle refers to the angle α which was used for the rotation of the second wafer compared to the first one during processing in the tool as shown in Fig. 2.

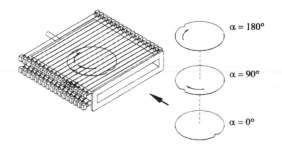

Fig. 2. Example of positioning of the wafers in the tool during processing.

$$T_1 = T, \tag{3c}$$

$$T_2 = \langle T_1 \rangle_R = \langle T \rangle_R. \tag{3d}$$

Here $\langle T \rangle_R$ stands for the rotation of the tool distribution by an angle $-\alpha$.

Now Eqs. (2a)–(b) can be simplified using Eqs. (3a)–(d):

$$X_1 = M + P + T + \Delta_1, \tag{4a}$$

$$X_2 = M + P + \langle T \rangle_R + \Delta_2. \tag{4b}$$

These equations are the basis for any further investigation of tool and preprocess uniformities.

3. Separation of tool uniformities

Using Eqs. (4a)–(b) it is possible to split the influences of superposed preprocess and tool influences in these distributions. The preprocess influence can be eliminated by a simple subtraction of both measured distributions,

$$X_1 - T_2 = T - \langle T \rangle_R + \Delta_1 - \Delta_2, \tag{5}$$

leading to a distribution which provides an overlay of the tool uniformity at two different angles. If the statistical deviations from the typical distributions are small compared to any other contributing distribution and if they have a random structure (as statistical deviations should have) they can be neglected.

$$T - \langle T \rangle_R \approx X_1 - X_2 \tag{6}$$

A specific solver algorithm based on matrix operations [2] was developed to solve this equation leading to the tool uniformity.

4. Separation of preprocess uniformities

The elimination of the tool influence in Eqs. (4a)–(b) is more complicated. Therefore the data set measured on the second wafer has to be rotated by the alignment angle α (see Fig. 2), which is expressed as $\langle X_2 \rangle_{-R}$.

$$\langle X_2 \rangle_{-R} = \langle M \rangle_{-R} + \langle P \rangle_{-R} + T + \langle \Delta_2 \rangle_{-R}. \tag{7}$$

Taking into account that the mean, M, is a homogeneous distribution results in

$$\langle X_2 \rangle_{-R} = M + \langle P \rangle_{-R} + T + \langle \Delta_2 \rangle_{-R}. \tag{8}$$

Subtracting this distribution from Eq. (4a) leads to the intended elimination of the tool influence

$$X_1 - \langle X_2 \rangle_{-R} = P - \langle P \rangle_{-R} + \Delta_1 - \langle \Delta_2 \rangle_{-R}. \tag{9}$$

If the difference of the two statistical deviations from the distributions is small enough it can be neglected.

Thus the overlay of the preprocess influence at two different angles can be expressed as

$$P - \langle P \rangle_{-R} \approx X_1 - \langle X_2 \rangle_{-R}. \tag{10}$$

The preprocess influence on the measured distribution can be calculated using a specific solver algorithm based on matrix operations [2].

5. Tests of the separation methods and substraction method

The separation methods were tested first with some artificially superposed preprocess and tool nonuniformities. These tests should show the ability of the algorithms to reproduce these (known) distributions. Three different tests were done. First two different gradients were superposed as preprocess and tool uniformities. The algorithms could determine both gradients without any problems.

Afterwards the separation methods were tested with a superposition of a gradient distribution as preprocess uniformity and a rotationally symmetric one as tool uniformity. The gradient distribution again could be reproduced without any problem. As expected from the applied method, the rotationally symmetric distribution could not be reproduced, because of loss of information in ring segments during subtraction of rotated data sets.

This can be overcome by a subtraction method which was developed especially to solve the problems with rotationally symmetric distributions. As long as either the process or the preprocess distribution shows non-rotationally symmetric features this subtraction method enables the user to split between both distributions. The only case which cannot be handled by the presented separation and subtraction methods is an overlay of both rotationally symmetric preprocess and process distributions.

In the following example the preprocess distribution P calculated by the separation method is subtracted from the measured data set (Eq. (4b)). Afterwards the resulting distribution is rotated by the alignment angle α,

$$\langle X_2 - P \rangle_{-R} = M + T + \langle \Delta_2 \rangle_{-R}. \tag{11}$$

The resulting distribution has the advantage of including the mean distribution showing the real dimensions of the overall values as well as the tool uniformity. In the case of rotationally symmetric preprocess uniformities the same method can be applied. Of course this subtraction method can also be used for any other kind of uniformity, if the real dimensions of mean values and deviations should be compared.

The third test was done superposing a real world tool distribution with an assumed gradient as preprocess uniformity under different rotation angles α. The results showed a very good reproduction of the gradient and an acceptable reproduction of the tool uniformity. An even better comparison to the original data sets could be found if the subtraction method is used to derive the tool uniformity from the superposed 'measurement' distributions and the calculated preprocess uniformity. This test was done for various

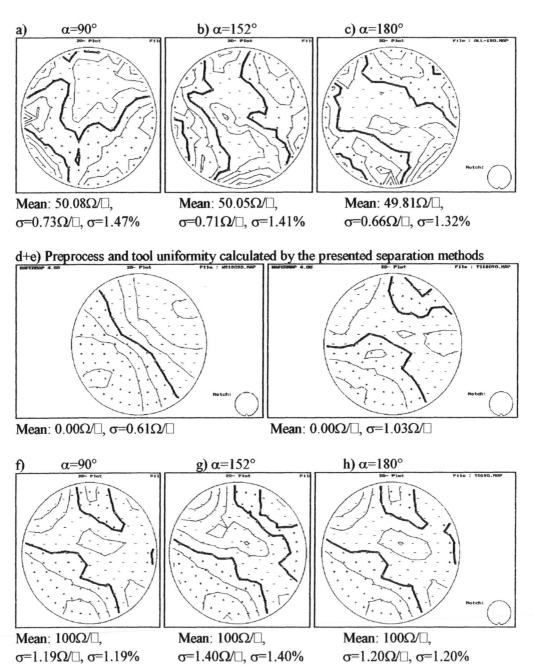

Fig. 3. (a–c) Sheet resistance distributions measured after implantation of 200 mm wafers (As$^+$, 30 keV, 1E16 cm^{-2}) and subsequent RTP annealing at 1050°C for 30 s soak time at different alignment angles α. (d and e) Preprocess (implanter) and (RTP) tool uniformities calculated by the presented separation methods using the distributions given in (a) and (c). (f–h) Tool uniformities calculated by the presented subtraction method for different wafer alignment angles α after normalization of the measurement distributions using the calculated preprocess distribution (d).

differences in the alignment angles α. It could be shown that small differences in the alignment angles should be favoured for the separation methods. Rotations as large as $\frac{1}{2}$, respectively 180°, lead to unnecessary loss of information for the solver algorithms. This loss of information is based in the mathematical problem and can not be overcome.

6. Application of the separation methods to experimental data

The presented separation methods were applied to sheet resistance measurements after implantation (As^+, 30 keV, 1E16 cm^{-2}) of 200 mm wafers. These wafers were processed at different alignment angles in an AST 2800ε RTP system [3] at 1050°C for 30 s soak time. The measured distributions are shown in Fig. 3(a)–(c). Because the mean values of the distributions, as usual in wafer processing, showed slight differences (49.81–50.08 $\Omega/\square$) all distributions were normalized to a mean of 100 $\Omega/\square$. The presented separation methods were applied to the normalized 90° and 180° distributions for calculation of the preprocess and tool uniformities. The results of these separations are given in Fig. 3(d) and (e). Fig. 3(d) shows a typical pattern of a batch desh implanter with a 45° scan direction. Fig. 3(e) illustrates the uniformity of the RTP tool.

Additionally, for testing of the integrity of the solutions the subtraction method was applied to the measurement distributions. The preprocess uniformity calculated by the separation method (Fig. 3d) was subtracted from the measurement distributions (Fig. 3a–c) leading to distributions including the overall mean and the tool uniformity (see Eq. (11)). First the subtraction method was applied to the 90° and 180° distributions (Fig. 3a and c). These results are shown in Fig. 3(f) and (h), respectively. It is not really surprising that these distributions are about the same. The reason for this is that the preprocess distribution was calculated from these two measurement distributions. If the theory and implementation of the algorithm are correct the substraction method applied to both wafers has to lead to the same results.

Afterwards, as a cross checking test, the subtraction method was applied to the 152° measurement distribution. The result of this operation could not be predicted, because the 152° measurement distribution was not used in the previous separation process of the preprocess and tool uniformities. The resulting distribution (see Fig. 3g) is very close to the previously calculated 90° and 180° ones (Fig. 3f and h). This agreement of all three distributions proves the presented methods.

Until now implanter monitoring [4] was only possible without separation of RTP and implanter nonuniformities. For future technology and applications [5] the knowledge of the status, uniformity and reliability of each single tool is indispensable. The presented methods allow for the separation and identification of single tool drifts, e.g. in statistical process control (SPC), as well as for uniformity identification of single tools during the development process.

7. Conclusions

Different methods were presented which allow for separation of input (preprocess) and process uniformities from resultant output distributions. These methods were applied to a specific problem in the semiconductor industry. Measurement distributions of sheet resistance after rapid thermal annealing processes were analysed for contributions from implanter and contributions from the RTP tool. It could be shown that the presented separation methods enabled the user to split up both uniformity contributions from the resultant measured parameter distribution on the wafer. The method can be used for daily monitoring of single tool drifts, which currently is not possible in some cases.

An application of the separation methods requires only two wafers preprocessed under the same conditions. These wafers have to be processed with same recipes but at different alignment angles in the investigated tool. The separation methods were tested under different conditions. They have shown their functionality with artificial distributions and real world data. The methods were tested for a variety of rotation angles. Rotation of the wafer by 180° before processing in the tool was identified as unfavourable configuration. Smaller rotation angles (e.g. 45°, 60° or 90°) should be favoured. Limitations of the method were investigated. One specific limitation is that rotational symmetric distributions cannot be reproduced. This would exclude the investigation of tools with wafer rotation from the presented methods, but a substraction method was developed to overcome this problem. This add-on-method can be used for all other cases too.

Acknowledgements

The help of Dr. Rolf Bremensdorfer, STEAG AST Elektronik, in carrying out the experiments presented in this paper is gratefully acknowledged. Furthermore the authors thank Varian IIS for supplying the test wafers. Special thanks to Jeff Gelpey and Christian Grunwald, STEAG AST Elektronik, for their help in preparation of this paper.

References

[1] WAFERMAP for Win95/NT 1.0. Users manual. Boin (GmbH), 1998.

[2] Boin M. Methods of separation of input and process distributions from measurement data sets. Internal report. Manuela Boin Scientific Software, 1998.

[3] Nenyei Z, Walk H, Knarr T. Defect guarded rapid thermal processing. J Electrochem Soc 1993;140(6):1728–33.

[4] Lerch W. Ion implantation and rapid thermal annealing in synergy for shallow junction formation. Phys Status Solidi (a) 1996;158:117–36.

[5] Nenyei Z, Wein G, Lerch W, Grunwald C, Gelpey J, Wallmüller S. RTP development requirements. In: Proceedings of the 5th International Conf. on Advanced Thermal Processing of Semiconductors RTP'97. New Orleans/Louisiana, 1997. p. 35–43.

PERGAMON

Materials Science in Semiconductor Processing 1 (1998) 201–205

MATERIALS
SCIENCE IN
SEMICONDUCTOR
PROCESSING

Coupled simulation of gas flow and heat transfer in an RTP-system with rotating wafer

S. Poscher [a],*, T. Theiler [b]

[a] Bereich Bauelementetechnologie, Fraunhofer Institut für Integrierte Schaltungen, Schottkystrasse 10, 91058 Erlangen, Germany
[b] Steag-AST-Elektronik, Daimlerstrasse 10, 89160 Dornstadt, Germany

Abstract

In this study the concept of a lamp heated RTP-system with rotating wafer is considered. Using the fluid-flow-simulation software Phoenics-CVD, we investigated the cooling of the wafer by a process gas flow which is injected at room temperature into the hot process chamber through an inlet pipe in the side wall. In a full 3D-simulation of the gas flow and of the heat transfer in the gas and in the wafer the Navier–Stokes equations and the energy equation are solved. The radiative power consumption and the energy loss of the wafer have been modeled by the Stefan–Boltzmann law. Simulations without wafer rotation show a strong drop in the temperature distribution at the wafer near the inlet pipe. In contrast to this, simulations with rotation show an axisymmetric temperature distribution with a considerably smaller temperature gradient over the wafer. Comparisons with oxidation experiments showed good agreement with the simulation results. © 1999 Elsevier Science Ltd. All rights reserved.

1. Introduction

Faced with shrinking device dimension and increasing wafer diameters in IC production, development of new rapid-thermal-processing (RTP) systems has to satisfy increasing requirements on temperature uniformity. Decreasing development cycles and increasing equipment costs require new methods before starting to build up new systems. One of these methods is equipment simulation. This example shows how wafer rotation influences the results of processes. The simulation results are compared with experimental results gained in a test apparatus.

2. Reactor concept

An experimental heating system has been built up with a quartz glass cylinder as process chamber (Fig. 1). The inner height of the chamber was 2.5 cm and the radius was 13.75 cm. The 200 mm silicon wafer was placed in the center of the chamber on quartz pins and was able to rotate during the process. The silicon wafer was surrounded by a 1 cm wide so called guard-ring made of the same material and thickness as the wafer. The development of conventional RTP-systems has shown that the use of a guard-ring results in a better temperature homogeneity at the wafer edge. The wafer was heated by two halogen lamp fields mounted above and below the process chamber. The process chamber was cooled by an external air flow, entering through small holes between the lamps. During the process, the chamber was purged by an internal process gas flow which enters through a gas inlet pipe (diameter 8 mm) in the side wall and leaves through a circular gap near the bottom. Gas distribution plates in front of the inlet tube often are used to avoid an one-sided temperature drop on the wafer caused by direct impinging of the jet. This study was performed without the use of plates because our main goal was to produce inhomogeneities in order to calibrate the simulator against the experiment.

* Corresponding author.

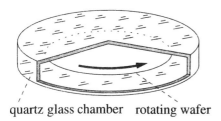

quartz glass chamber   rotating wafer

Fig. 1. Schematic drawing of the process chamber.

3. Equipment simulations

The gas flow and the heat transfer inside the process chamber were simulated using the computational-fluid-dynamics (CFD) software Phoenics-CVD [1]. A 3D-simulation model was developed, describing the gas room, the silicon wafer and the guard-ring. The pins were neglected due to their low radius. The simulation area was meshed by a cylindrical polar grid. Velocity, pressure and temperature distributions were computed by solving the steady state Navier–Stokes equations and the energy equation numerically. The outlet was set to atmospheric pressure. The density of the gas was calculated from the ideal gas law, viscosity and thermal conductivity were calculated from kinetic gas theory in dependence from the local gas temperature. The process gas was assumed to be pure oxygen as used in rapid thermal oxidation. (Since viscosity and conductivity of nitrogen differs from this only by about 15%, simulation results are also applicable to implant anneal processes, where nitrogen is used as process gas.) The thermal conductivity of the wafer and of the guard-ring was assumed to be constant and the value reported in literature for silicon at 1400 K was used [2].

Since the reactor has a very small height, Rayleigh–Benard convection rolls are not expected. Therefore for the purpose of faster convergence the gravity term in the Navier–Stokes equation was neglected. Furthermore the Reynolds number stays below the critical value for the occurrence of turbulence. Therefore turbulence models, which describe eddies smaller than the grid cells, need not to be considered.

The inlet temperature of the gas was set to room temperature, the chamber walls were set to $T_{wall} = 700$ K. At the wafer and at the guard ring the radiative heat consumption per area q was set by the following equation:

$$q = \sigma \epsilon (T_{wafer}^4 - T_0^4) \tag{1}$$

where $\sigma = 5.67 \cdot 10^{-8}$ W/m K^4 is the Stefan–Boltzmann constant, $\epsilon = 0.7$ is the emissivity of silicon and the value of the constant $T_0 = 1412.0$ K has been chosen in a way, that a maximum temperature on the wafer of

$T_{wafer,max} = 1400$ K is achieved. Eq. (1) is derived from the Stefan–Boltzmann law and describes the heat reception of an infinite plate which is irradiated by an infinite lamp field with homogeneous temperature and power density. During the simulation the local temperature on the wafer T_{wafer} is calculated from the interaction of radiative heat consumption and convective heat removal.

As usual in laminar flow simulations, the gas velocity at the chamber walls was set to zero. The angular velocity of the wafer and of the gas at the wafer surface was given by the number of revolutions per second.

The differential equations were solved by the finite volume algorithm. Achieving convergence needed about 4000 iterations. Convergence is achieved if temperature and velocity fields do not change any more at further iterations, if inflow and outflow differ less than 0.1% and if the sum of the absolute error of the finite volume equation summed over all grid cells has dropped to 10^{-4} times the initial value [3]. Several grid refinements were executed, until temperature isolines interpolated by the graphic postprocessor did not change at further refinements. Finally a grid of 39 cells in tangential direction, 22 in radial and 16 in axial was found to be sufficient. The grid resolution must be highest near the inlet jet where the maximum gradients appear. One simulation needs about 8 h on a DEC-Alpha-Server 2100 5/250.

4. Results and discussion

Figs. 2 and 3 show the calculated gas flow in the process chamber at a flow rate of $F = 5$ slm and an assumed maximum wafer temperature of 1400 K, if the wafer does not rotate. The gas jet enters the chamber in all figures from the upper left side. It can be seen from the streamlines that the inlet jet spreads over the whole wafer diameter. In a simulation performed at room temperature for comparison, essentially less jet spreading appears. In this case the streamlines are more dense and spread only after impinging at the opposite wall. The faster spreading of the inlet jet at high temperature is caused by the strong dependence of viscosity from the temperature: $v \sim T^{1.7}$. In this way it could be demonstrated that simulation at process temperature can visualize effects that can hardly be seen in experiments near room temperature.

The one-sided exposure of the wafer to the gas jet leads to a one-sided temperature drop at the wafer of $\Delta T = 60$ K (Fig. 4). At the guard-ring even a on-sided temperature drop of $\Delta T = 70$ K has been computed (not shown in Fig. 4). For experimental verification of

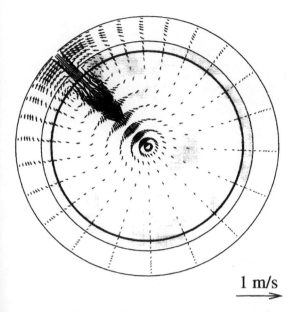

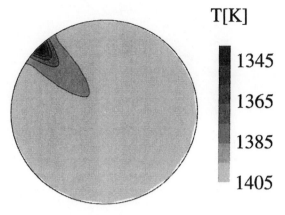

Fig. 4. Simulated temperature distribution on the non-rotating wafer.

1 m/s

Fig. 2. Simulated velocity distribution between wafer and chamber ceiling without wafer rotation.

these simulation results, an oxidation on a 100-Si-wafer for 3 min has been performed. Fig. 5 shows the layer thickness measured by an ellipsometer after the process. High layer thickness corresponds to high temperature and vice versa. There is good qualitative agreement. The predicted one-sided temperature drop can clearly be seen in the experiment.

Fig. 6 and Fig. 7 show the computed gas flow at the same process conditions, when the wafer rotates with a frequency of $f = 1$ s^{-1}. Wall friction causes the gas to follow the wafer rotation particularly before leaving the chamber. The calculated temperature distribution is axisymmetric for this case (Fig. 8). Since cooling by the inlet flow is distributed over the whole wafer edge, a significant lower temperature drop over the wafer of about $\Delta T = 3.3$ K results. That way it has been shown that much better homogeneity can be achieved with a rotating wafer than with a static wafer, assuming that in both cases no gas distribution plate is used. Again Fig. 9 shows the oxide layer thickness measurement at the same process conditions. The distribution is not completely axisymmetric as expected. This might be due to oscillations of the lamp control caused by a

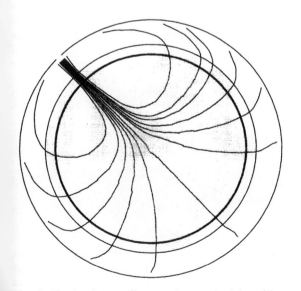

Fig. 3. Simulated streamlines starting at the inlet without wafer rotation.

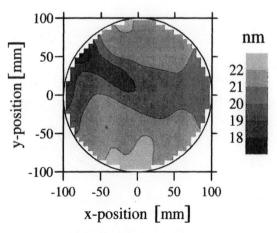

Fig. 5. Experimental oxide thickness at non-rotating wafer.

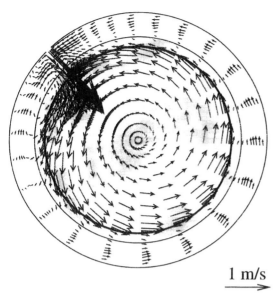

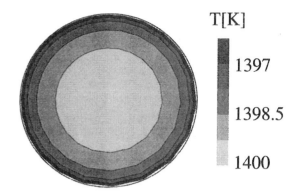

Fig. 8. Simulated temperature distribution at the rotating wafer.

1 m/s

Fig. 6. Simulated velocity distribution between wafer and chamber ceiling with rotating wafer.

Experiments and simulations with rotating and static wafers have been executed at various gas flow rates. Simulated maximum temperature differences on the wafer are compared to those calculated from experimental oxide thicknesses in Fig. 10 (3 mm edge exclusion). For the static case, the curve is predicted qualitatively correctly, although the increase is too steep in the simulation. In the rotating case, the experimental temperature drops are clearly higher than the simulated ones. This is probably due to geometric and spectral aspects of radiation which are not considered in the simple formula in Eq. (1). However trends can be very well predicted with the existing model.

rotating thermocouple in the experiment. This effect was not considered in the simulation.

For a quantitative verification, the following relationship between oxide thickness drop Δd and temperature drop ΔT [4] has been used:

$$\frac{\Delta d}{\Delta T}\bigg|_{1400 \text{ K, 3 min, 1 atm, 100-wafer}} = 0.16 \text{ nm/K} \qquad (2)$$

5. Conclusions

The study demonstrates that simulation is an appropriate method to promote development of semiconductor equipment. Physical effects can be visualized and

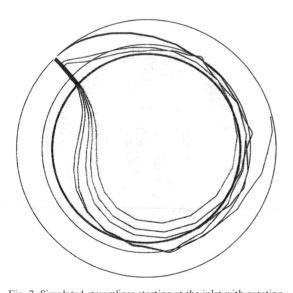

Fig. 7. Simulated streamlines starting at the inlet with rotating wafer.

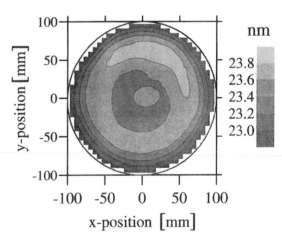

Fig. 9. Experimental oxide thickness at the rotating wafer.

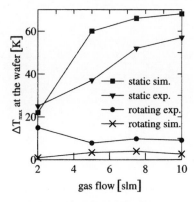

Fig. 10. Maximum temperature difference on rotating and non-rotating wafer at various gas flows. Comparison of simulation and experiment.

their influence on uniformity can be quantitatively estimated. The achieved accuracy of the simulations fully satisfies the requirements specified in the 1997 edition of the SIA Road-map [5]. Experiments and simulations show that wafer rotation can reduce several kinds of inhomogeneities which could appear on non-rotating wafers.

References

[1] Spalding B, editor. The Phoenics journal of computational fluid dynamic and its applications, vol. 8, No. 4. Wimbledon, December 1995.

[2] Electronic Materials Informations Service: EMIS data-reviews series. Properties of silicon. London: Inspec, 1988.

[3] Ludwig JC. Phoenics-2 instruction course notes. CHAM TR300. Wimbledon, 1994.

[4] Zhi-min Ling. Silicon oxidation: modeling, characterization and strategic considerations. Ph.D. thesis, Leuven, 1989. p. 181.

[5] The national technology road-map for semiconductors. 1997 ed. Austin: SEMATECH, 1997. p. 190.

PERGAMON

Materials Science in Semiconductor Processing 1 (1998) 207–218

MATERIALS
SCIENCE IN
SEMICONDUCTOR
PROCESSING

Dopant diffusion studies and free carrier lifetimes during rapid thermal processing of semiconductors

Rajan V. Nagabushnam[a],*, Rajiv K. Singh[a], Sujit Sharan[b]

[a]*Department of Materials Science and Engineering, University of Florida, Gainesville, FL 32603, USA*
[b]*Process Development, Micron Semiconductor, Boise, ID 83705, USA*

Abstract

Rapid thermal processing of semiconductors involves significant photonic and subsequent thermal excitation. In the past, photonic excitation during rapid thermal annealing had been speculated to lead to significant enhancement of dopant diffusion or activation. In this work we present some experimental results indicating the absence of any such enhancement at high temperatures (~1000–1050°C) which most often are employed during the metal-oxide–semiconductor device processing. The implanted dopant (boron, arsenic or phosphorus) movement in silicon during different rapid thermal annealing conditions was studied using secondary ion mass spectroscopy (SIMS) technique. To understand the effect of point defects in controlling the diffusion process, the concentrations of charged and neutral point defects were calculated as a function of carrier concentration using previously published defect-carrier relations. The dependence of free carrier concentration on lattice perturbation parameters such as impurities and temperature was formulated and used in calculating carrier lifetimes (τ) in silicon. We qualitatively analyze two competing reactions, (i) the phonon release at the defect sites and (ii) the Auger electron process due to many electron interactions, to explain the apparent absence of any enhanced dopant diffusion. In our analyses, we obtain a highest free carrier lifetime of about 442 ns in the case of low dose (1e13/cm^2) implanted sample during the transient stage (700°C) of the dopant activation cycle. The corresponding smallest (~17 fs) free carrier lifetime was obtained for the high dose implanted sample (dopants already activated) at 1000°C, the steady state part of an extended anneal cycle. Based on the detailed free carrier lifetime analyses, we suggest that any enhanced dopant activation or diffusion, at the best, may occur only at very low temperatures in the samples implanted with low doses of dopant atoms. © 1999 Published by Elsevier Science Ltd. All rights reserved.

1. Introduction

The electron–hole recombination process, characteristic of semiconductors, has found its way in a crucial area of VLSI device processing: rapid thermal processing (RTP) [1–13]. In the RTP method, the wafer is heated from either one side or both sides by intense lamps (as high as 25 W/cm^2) for short periods of time (< 60 s). The wafer sits on quartz pins or slip-free ring and receives the radiation from the lamps through a quartz window which is transparent to infrared radiation. This single wafer processing technology can perform any process requiring thermal energy, such as oxidation, nitridation, silicidation, implant anneal, BPSG reflow and chemical vapor deposition [7, 8, 11]. As the device size becomes smaller (sub 0.25 μm), the very large scale integrated (VLSI) industries are beginning to replace furnace based processes with RTP sequences, where the thermal budgets are significantly lower. The heating sources generally comprise of tungsten halogen lamps in which the filament temperature reaches as high as 3500 K. A significant portion of this radiation consists of ultraviolet and infrared photons with photon energies greater than the silicon band gap (~1.12 eV).

* Corresponding author. Present address: Networking and Computing Systems Group, Motorola, Austin, TX-78731, USA.

Though successful fabrication of sub 0.25 μm devices with source/drain implant anneals performed using RTA has been demonstrated many times [12, 13], there has been a lack of studies aimed at understanding the role of photons in dopant diffusion during RTP. It has been commonly observed that the movement of certain dopants, such as boron and phosphorus, during rapid thermal activation anneal is more than what is estimated using conventional equilibrium point-defect based models. This phenomenon is commonly called 'transient enhanced diffusion (TED)' supposedly arising out of implantation damage and has been an active subject of research for several researchers in the past decade [14–16]. However, some researchers have suggested that the high energy photons, in addition to TED, present exclusively during rapid thermal processes, either generate excess point defects in silicon or modulate the charged to neutral point defect ratio both of which could potentially also lead to enhanced activation/diffusion of dopant atoms [17–21].

Though the above mentioned arguments had been based on certain observed dopant concentration profiles or silicon surface sheet resistance values, the questions regarding accuracy and reproducibility of the sample temperature always existed. It has been well known that an accurate temperature control has been the biggest challenge faced by various commercial RTP tools before gaining any acceptance into the manufacturing environment. In spite of several years of experimentation with various temperature measurement techniques ranging from laser interferometry to acoustic methods, the *non-contact* infrared pyrometry and the *contact or near contact* thermocouple measurements have remained as the most reliable measurement tools in practice [7]. It is to be remembered that both of these techniques, employed in the conventional manner, also have their limitations. The pyrometry technique, in general, employs a sensor, usually located beneath the wafer, which supposedly receives the radiation only from the wafer and uses it to calculate the wafer temperature. The accuracy of the calculated temperature primarily depends on the accuracy of one programmed input parameter 'emissivity of the wafer' [7–10]. The emissivity, in turn, is not only a strong function of the layers present on the wafer, doping condition and surface roughness of the wafer, but also varies with temperature. Hence it is difficult and complicated, in practice, to accurately estimate the emissivity of the wafer as some of the required input parameters have not been fully quantified yet.

In addition, if the wafer is partially transmissive, the emissivity (ε) will be a function of absorptivity (α) of the wafer. The absorption characteristics of the incident electromagnetic radiation is a strong function of the free electron concentration and band gap of the substrate material, when the overlying layers are too thin to absorb the incident radiation completely. The absorptivity is also dependent on substrate temperature which determines not only the free electron concentration but also the phonon (transverse and longitudinal) density. Thus, wafers having different free electron concentrations can significantly affect the emissivity especially at lower temperature ($< 7000°C$). Seidel et al. reported that N^+ doped wafers ramped up at a rate twice than N^- doped wafers for the same input power conditions [9]. Thus the data referring to anomalous dopant diffusion or activation obtained by previous researchers may be partially due to the lack of accurate temperature control in these experiments. Thus, in practice, it would be difficult to subject two different wafers having varying different emissivities to the exact same thermal cycle, where the temperature controller is based on infrared pyrometry. On the other hand, the thermocouple based measurement technique suffers from the possibility of poor thermal contact, which could degrade due to stresses developed during processing.

In this research, we designed a new set of diffusion experiments without any of the deficiencies related to the accurate and reproducible temperature control. This was accomplished by employing a wafer emissivity independent temperature measurement technique. In addition by annealing the implanted wafers in two different configurations (i) implanted side facing the lamps and (ii) wafer back side facing the lamps, otherwise exact conditions were maintained and photonic effects estimated. A detailed theoretical analysis of the factors that can affect dopant diffusion in silicon has also been investigated.

2. Experiment

A rapid thermal system (Applied Materials Centura[R]) using a wafer emissivity independent pyrometry was used in these studies. The radiation cavity surrounding the pyrometer is designed to maintain an effective emissivity of unity under all conditions. This system, when operated in a closed loop mode (temperature control), thus ensures that two different wafers with varying emissivities will be subjected to an identical thermal cycle. The heating lamps are located only on the front side of the wafer. The 6 inch wafer sits on the slip-free ring made of polysilicon and rotates during processing. Thus by employing this arrangement, the temperature of the wafer was determined independent of its absorptivity and emissivity. In addition, this system was able to deliver excellent thermal uniformity over the 6 inch wafer.

P-type (1×10^{13} background doping) (100) oriented CZ silicon wafers were used in all of our studies.

Table 1
Various implantation conditions

S. No.	Species	Screen oxide (Å) before implantation + before annealing	Low dose (/cm^2), implantation energy (keV) (LD)	High dose (/cm^2) implantation energy (keV) (HD)
1	arsenic	200 Å thermal + 200 Å PECVD oxide	1×10^{13}, 70	5×10^{15}, 70
2	boron	200 Å thermal	1×10^{13}, 50	5×10^{15}, 70
3	phosphorus	200 Å thermal + 200 Å PECVD oxide	1×10^{13}, 70	5×10^{15}, 70

Following the growth of a 20 nm thick screen oxide by a furnace process, the wafers were implanted with one of the three dopants: phosphorus, boron or arsenic. Two different types of implantation conditions (1×10^{13} or 5×10^{15}/cm^2) were employed for our studies. The complete details relating to dose and implantation energies are shown in Table 1. The low dose rates were chosen to avoid the formation of extended defects and their effect on dopant movement during annealing. In addition, a 20 nm thick PECVD silicon dioxide was deposited at 400°C before the annealing, on the wafers implanted with either phosphorus or arsenic (capping layer) to minimize any dopant loss during high temperature annealing.

The wafers were subjected to either one of the two types of thermal cycles. As shown in Fig. 1, the first cycle (type I) involves heating the wafer, at the rate of 35°C/s, to the set point anneal temperature (1000°C in the case of phosphorous or boron implanted wafers and 1050°C for arsenic implanted wafers) with immediate cool down to room temperature. In the second case (type II), the wafer was held at the high temperature for a short duration (20 s) and then brought

down to room temperature. In both thermal cycles, the wafer is held for a period of 30 s at 750°C to enable improved temperature uniformity across the wafer during the high temperature anneal step. These two different thermal cycles were chosen to investigate the possible influence of photons on dopant activation/diffusion during different stages (transient versus steady state) of a rapid thermal cycle.

The actual experiments involved subjecting the wafers, with identical implants, to the above mentioned thermal cycles in two different configurations: (i) implanted side of the wafer facing the heating lamps and (ii) wafer backside irradiated by the lamps. In the first set of conditions, the dopant surface receives a high dose of photons which subsequently converts into thermal energy by their absorption by electrons whereas in the second case, the wafer backside receives the photonic flux with the dopant surface receiving the normal thermal excitation only. This helped us to separate the thermal and non-thermal factors contributing to the dopant diffusion/activation. Our calculations and other published literature also suggest that the temperature differential between the top and bottom surfaces of a 675 µm thick wafer is less than 1°C and thus the thermal effects are expected to be the same [10]. It should also be noted that though earlier experiments, conducted by others, could also have possibly involved such an experimental design in terms of whether the implanted surface was facing the heating lamps or not, the lack of emissivity independent temperature measurement technique might have lead to the misinterpretation of dopant diffusion or activation results. After the anneal, the screen oxide was stripped using a 1:10 HF–H$_2$O etchant mixture for sheet resistance measurements and secondary ion mass spectroscopy (SIMS) analysis. Sheet resistance measurement was carried out at 49 points across each wafer using the standard four point probe method.

3. Results and discussion

The measured sheet resistance (a variable dependent on both the amount of electrically active dopant species as well as the depth up to which they are present) values from each wafer are tabulated in Table 2.

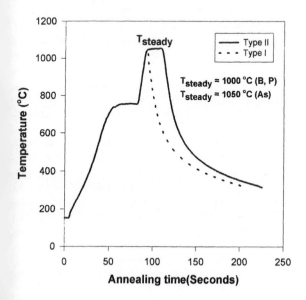

Fig. 1. Two types of annealing cycles used for our experiments.

Table 2
Measured sheet resistance values for different annealing conditions

S. No.	Dose type	Thermal cycle	Species	Implanted side illuminated ($\Omega/\square$)			Backside illuminated ($\Omega/\square$)			% Variation
				mean	S.D.	range	mean	S.D.	range	
1	low dose	type II	boron	205.37	0.63	5.0	201.18	0.73	6.4	2.04
2	low dose	type II	arsenic	2734.2	1.7	215	2728.3	1.81	210.0	0.21
3	low dose	type II	phosphorus	1880	0.69	55.0	1812.4	0.82	63.0	3.5
4	high dose	type I	boron	57	2.17	4.61	57.2	2.68	6.93	−0.35
5	high dose	type I	arsenic	70.8	0.84	2.57	71.1	0.9	2.8	−0.42
6	high dose	type I	phosphorus	28.8	0.36	0.43	29	0.25	0.28	−0.7
7	high dose	type II	boron	34.1	1.46	1.42	34	1.75	2.1	0.3
8	high dose	type II	arsenic	62	0.68	1.8	62.2	0.75	1.98	−0.3
9	high dose	type II	phosphorus	26.9	0.36	0.35	27.3	0.36	0.35	−1.4

The mean, standard deviation (3σ) and the range of the sheet measurements are shown in Table 2. Table 2 shows that, as expected, the sheet resistance decreases with longer annealing times (type II anneal) and with increasing dose. The standard deviation of the measured sheet resistance varied from 0.25 to 2.68%, thus confirming the excellent thermal uniformity across the wafer. Table 2 shows that the sheet resistance of the wafers, for a given implantation condition and thermal cycle, is the same whether the surface was facing the lamps or not. The difference between the corresponding mean sheet resistance measured on wafers with identical implantation conditions, but annealed in two different configurations, ranges from −1.4 to 2%, which are within the limits of experimental error. From these values, we can conclude that the total amount of electrically active dopant species was only dependent on the absolute conditions relating to implantation (dose and energy) and the thermal cycle (annealing set point and time) and not on the way the wafers were annealed, i.e. whether the implanted side was facing the lamps or not.

The SIMS data showing the dopant movement during various rapid thermal annealing conditions is shown in Figs. 2–4. Fig. 2 shows that boron concentration profiles (BSi[39] cluster monitored) after the annealing involving no dwell time at 1000°C resemble the as implanted profile. In addition, it could also be seen that the concentration profiles for the two types of annealing configuration, being studied, lie above each other. In the case of phosphorus samples, the atomic concentration profile was affected by the nature of the annealing cycle (Fig. 3). For example, the peak P concentration (5×10^{15}, 70 keV) in the as implanted condition is approximately $4\times10^{20}/cm^3$ whereas it is about 3×10^{20} and $2.5\times10^{20}/cm^3$ for the annealing conditions involving dwell times of '0' and '20' s, respectively (Fig. 3b and c). However in the case of low dose samples (1×10^{13}, 70 keV), no significant atomic move-

ment was observed. This could have been due to the limitation posed by the resolution (PSi[61] and P[31] clusters were monitored) of our SIMS setup or the lack of extensive damage with the usage of low dose leading to reduced or insignificant atomic diffusion. In the case of arsenic implanted samples, the peak concentration (cluster SiAs[104] monitored) decreased from 9×10^{20} to $7\times10^{20}/cm^3$ in the case of annealing involving '20 s' dwell time at 1050°C. However the most striking feature of all these profiles in the case of all the three dopants is that the atomic concentration profiles were only determined by the implantation conditions, the annealing temperature and the annealing time and not by the sample-annealing system configuration maintained during the annealing. From both the sheet resistance measurements and atomic concentration profiles estimated by SIMS, it is clear that only the thermal component of the photon excitation process controls the *short range* as well as the *long range* diffusion of the impurities. If this was not true, the atomic concentration profile for any of the implanted species (boron, phosphorus and arsenic) subjected to thermal cycle with the implanted side facing the lamp would be different (modulated short range or long range diffusion) compared to the one with wafer backside facing the lamps due to the photon induced diffusion effects.

To explain the absence of any direct role of photons on dopant diffusion in silicon other than thermal heating, it is essential to look at certain generation and recombination events in silicon during rapid thermal annealing.

The photons incident on silicon would excite (i) electrons from the valence band into conduction band and (ii) electrons from certain defect centers such as point defects, impurity atoms, dislocation, etc. The relative importance of these two events primarily depends on the incident photon energy as well as the relative population of these two free electron sources. Though the energies required to liberate the free carriers from the

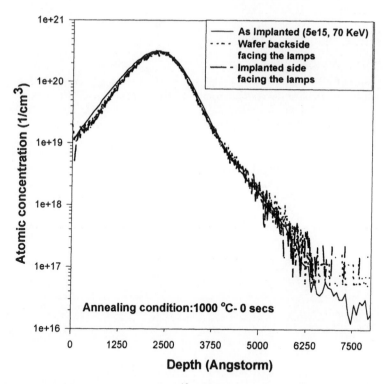

Fig. 2. Atomic concentration profile of boron (5×10^{15}, 70 keV), annealed 1000°C, 0 s, as measured by SIMS.

defect centers is lower than that required to excite the electron across the band gap at any temperature, it is fairly easy to realize that most of the free carriers making up the e–h plasma come from the sites in the valence band as the silicon lattice sites statistically outnumber the defect centers (typically about 10^4 to 10^6 times). At high temperatures, as the concentration of the free electrons increases, the incident photons would also be absorbed by the free electrons for the *intraband* transitions. As silicon is an indirect band gap material, the excitation would involve a phonon, which means that the probability of excitation of the free carriers becomes more probable at high temperatures as compared to lower temperatures. With a continued flux of photons, silicon is expected to be flooded with free carriers if recombination mechanisms do not exist. However as dictated by the law of mass action, the recombination mechanisms become more important as the density of non-equilibrium carriers increase. The recombination of the excited carriers would either result in the emission of thermal energy (phonon) or ejection of a free carrier (Auger process). If the recombination of free carriers takes place at a defect site such as a charged point defect, the defect could move either due to (i) the binding energy being locally released ('thermal kick out' mechanism) or (ii) the change of the charge state of the defect in itself ('Bourgoin' mechanism) [1, 2].

It is well known that, in addition to the neutral point defects, point defects of different charge types are present in silicon. The charged defects can be assumed to evolve as follows [22]:

$$X^0 + e = X^- \quad \text{or} \quad x^0 = X^- + h,$$
$$X^- + e = X^{2-} \quad \text{or} \quad X^- = X^{2-} + h,$$
$$X^0 + h = X^+ \quad \text{or} \quad X^0 = X^+ + e,$$
$$X^+ + h = X^{2+} \quad \text{or} \quad X^+ = X^{2+} + e \qquad (1)$$

where X^0, X^-, X^{2-}, X^+ and X^{2+} refer to neutral, single negative, double negative, single positive and double positive charged point defects (vacancy or interstitial), respectively. Though these point defects interact with free carriers (n and p) to transition from one type to the other, *dynamic equilibrium*, in which the concentration of each type of defect remains relatively constant with respect to each other and to the total concentration of free carriers, is always maintained. Hence any change in absolute free carrier concentration would change the absolute concentration of various types of point defects. It is to be remembered that the relations relating to concentration of charged defects as formulated by Shockley and Last [23] only include the intrinsic properties of silicon and thermodynamic parameter, temperature (T). It does not take into account the interaction of the defects with any excess free carriers (n, p), generated through external fields such as optical injection (lasers, RTA, etc.) or electric or electromagnetic fields.

In the 1970's, Watkins and coworkers [2] inferred the most acceptable values for the activation enthalpy associated with normal thermal migration, ΔH_M, from the annealing kinetics of defects, of the single vacancy (V) in Si in three of its five charge states. Using a similar annealing kinetics study approach [24], migration enthalpy of various AV pairs were also calculated by Hirata et al. These values are summarized in Table 3. It is fair to infer from the listed values that the corre-

sponding values of enthalpy of migration associated with singly charged vacancy would lie somewhere between that of the neutral and the doubly charged vacancy. Table 3 also indicates that the charged pair (PV^-) has a higher enthalpy of migration compared to the neutral pair (PV^0). Therefore any increase in (V^-) concentration due to the conversion of (V^0) to (V^-) would increase the overall enthalpy of migration, $\Delta H_M(PV)$ of phosphorus and hence the activation energy of phosphorus diffusion process, a part taking place with the aid of vacancy population. This argument regarding point defect charge conversion and its

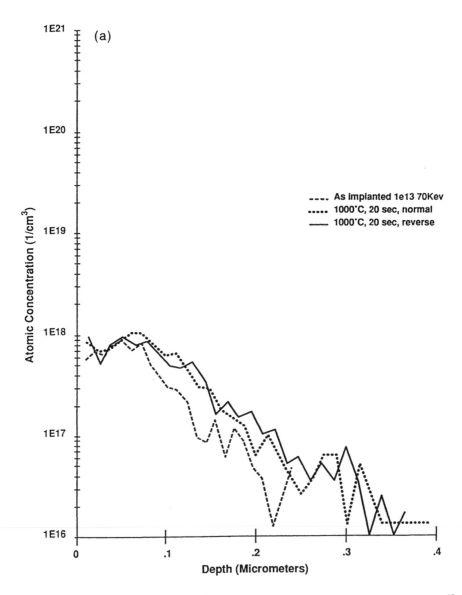

Fig. 3. Atomic concentration profile of phosphorus $(5 \times 10^{15}, 70 \text{ keV})$ as measured by SIMS. (a) Low dose $(1 \times 10^{13}, 70 \text{ keV})$, type I $(1000^\circ\text{C}, 0 \text{ s})$ condition. (b) High dose $(5 \times 10^{15}, 70 \text{ keV})$, type I $(1000^\circ\text{C}, 0 \text{ s})$ condition. (c) High dose $(5 \times 10^{15}, 70 \text{ keV})$, type II $(1000^\circ\text{C}, 20 \text{ s})$ condition.

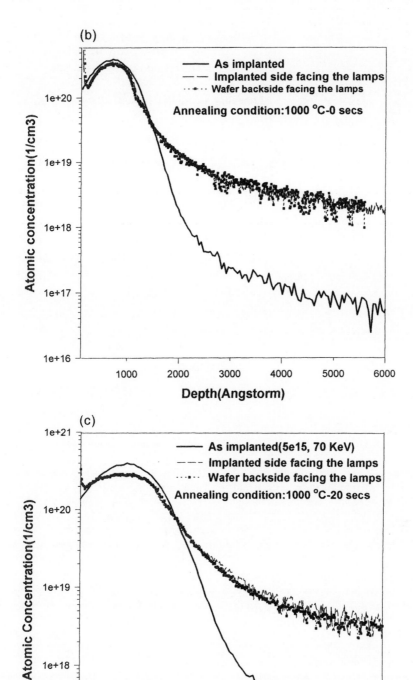

Fig. 3 (continued).

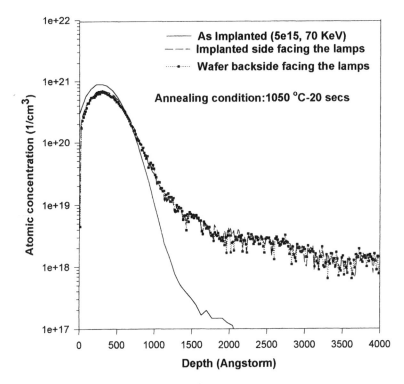

Fig. 4. Atomic concentration profile of arsenic (5×10^{15}, 70 keV, annealed 1050°C, 20 s, as measured by SIMS.

impact on diffusion is equally valid in the case of interstitials.

In addition, the enthalpy released [25] due to *electron–hole* annihilation at any temperature would depend on the site at which annihilation occurs and nature of the binding between the recombination site and the free carriers, i.e.

$$\Delta H_{\text{Rec}} = \Delta H_{\text{C–V}} - H_{\text{Bind}} \tag{2}$$

where $\Delta H_{\text{C–V}}$ and H_{Bind} refer to the enthalpy corresponding to the band gap and binding energy of the free carrier(s) to the recombination site, respectively. H_{Bind} could also said to be an indication of the depth of ionization level of the defect center (vacancy, interstitial, impurity–point defect complex, dislocations, stacking fault, etc.). The temperature dependence of binding energies for vacancies of various charge types had been discussed by Van Vecten extensively in his earlier work [26]. The binding energies for various defects, as suggested by Van Vecten, is summarized in Table 4.

As Table 4 shows, the exciton–point defect pair has the lowest binding energy (0.015 eV) followed by the free carrier–point defect and charged point defect–impurity atom pairs [5]. As the published literature on the various ionization levels of point defects deals pri-

Table 3
Migration enthalpy of point defects/dopant-point complexes

Point defect/dopant-point defect pair	Migration enthalpy, ΔH_{M} (eV)	Temperature range (K)
V^{2-}	0.18 ± 0.02	70–85
V^{0}	0.45 ± 0.02	220
V^{2+}	0.32 ± 0.02	150–180
PV^{0}	0.94	423–523
PV^{-}	1.25	423–523
AsV^{0}	1.07	423–523
SbV^{0}	1.28	423–523
BiV^{0}	1.46	423–523

Table 4
Binding energies of different point defect–dopant complexes

Point defect or point defect– dopant configuration	Binding energy (eV)	Temperature range (K)
Exciton–point defect pair	0.015	
$Hi(V^+)$–H_V	0.05	$T < 100$
$Hi(V^{2+})$–H_V	0.13	$T < 100$
H_C–$H_i(V^-)$	0.65	$T < 100$
H_C–$H_i(V^{2-})$	0.29	$T < 100$
$E_{Phosphorus-Vacancy}$	1.04	413–433
$E_{Arsenic-Vacancy}$	1.23	413–433
$E_{Antimony-Vacancy}$	1.44	413–433
$E_{Bismuth-Vacancy}$	1.64	413–433
$E_{Phosphorus-Interstitial}$	1.49	
$E_{Bismuth-Interstitial}$	1.52	
$E_{Boron-Vacancy}$	0.9	

marily with vacancies in silicon [27], we would restrict our discussion to vacancies. As H_{Bind} for the exciton pair is fairly independent of the temperature, ΔH_{Rec} would vary from 1.158 to 1.295 eV for temperatures ranging from 0 to 1323 K. This energy, released locally in appropriate mode ('kick' mechanism), is sufficient for all the above mentioned vacancies (listed in Table 3) to migrate with the aid of a little additional energy (which could come from thermal contribution), ΔH_{add}, less than 0.5 eV. Hence, according to this simplistic approach, the recombination of the excess free carriers at these defect sites should lead to an additional excess movement of the dopant atoms with only a very little increase of the wafer temperature during RTA. However, as there was no evidence of any enhanced dopant activation/diffusion in our studies due to high energy photons using RTA, it is worthwhile looking at other competing mechanisms, such as Auger processes, occurring in silicon which could partly explain our observation. As the excess free carrier concentration (Δn and Δp) induced by the photon irradiation is much higher than (n and p) we have focused on the recombination processes due to many electron and hole collisions.

All the Auger recombination processes fall into a few basic categories: (i) two phononless, (ii) two phonon-assisted band–band processes, (iii) four processes involving one type of trap and (iv) two donor–acceptor processes. Refs. [28–31] deal extensively with this topic. In general the rate of Auger process is given by Bn^2p, where B (having a unit of $cm^6\,T^{-1}$) refers to an Auger coefficient with n and p referring to the concentrations of majority and minority carriers, respectively. Hence as n increases, the probability of the reaction increases too. The unique characteristic of Auger processes, unlike any other processes, is the insignificant or null dependence of 'B' on temperature with the carrier lifetime being defined only by its concentration.

Based on a detailed derivation, one gets for the carrier lifetime (τ),

$$1/\tau = \alpha + \beta\Delta n + \gamma(\Delta n)^2 \tag{3}$$

where $\gamma = B_1 + B_2$, a widely accepted value of $3.79 \times 10^{-31}\ cm^6/s$. α and β are constants dependent on single electron transition and Auger process coefficients, excess free carrier concentration (Δn), modified carrier concentration (n_1) and trap density (N_t).

An experimental annealing cycle for our calculations regarding the free carrier lifetimes is shown in Fig. 5. Though this cycle is different from that used for the actual annealing, it would become clear at the end that we could only overestimate the excess free carrier lifetime by such a choice involving low levels of photon

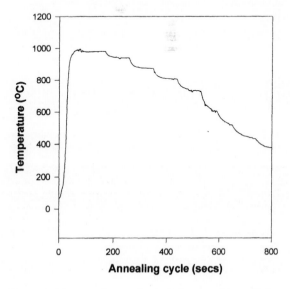

Fig. 5. RTA annealing cycle used for our Auger lifetime analysis.

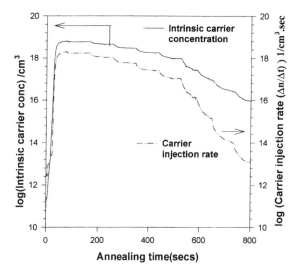

Fig. 6. Comparison of intrinsic carrier concentration and carrier injection rate for the RTA cycle under consideration.

irradiation. The intrinsic carrier injection rate due to thermal activation and the absorbed photon density in the top 700 Å of silicon is shown in Fig. 6. The absorbed photon density (dn), on the basis that the photons of energies as low as half the band gap of silicon are absorbed with the aid of phonons, is calculated using the Planck's radiation law and is given by

$$dn(\lambda, T) = \frac{\epsilon C_1}{\lambda^4(e^{C_2/\lambda T} - 1)} 6.24 \times 10^{18} \, d\lambda \, (\text{cm}^3 \, \text{s})^{-1} \quad (4)$$

where λ is the wavelength and the first and second radiation constants, C_1 and C_2, are equal to 3.7415×10^{-16} W m^2 and 1.43879×10^{-2} m K, respectively.

For a detailed carrier lifetime analyses, we would consider four different conditions, as shown in Table 5, for both types of implantation conditions used: (i) low dose (1×10^{13}, 70 keV) and (ii) high dose (1×10^{13}, 70 keV). Though this type of analyses could be performed for silicon implanted with either an n-type or p-type dopant, for simplistic purposes, we would restrict our analyses to silicon implanted with n-type dopant, say arsenic or phosphorus. The choice of the dopant determines the dominant trap responsible for the free carrier recombination. Conditions I and III physically represent the transient and steady state stages of a typical source/drain implantation anneal cycle, whereas the conditions II and IV represent the same corresponding stages of a thermal cycle used, say after the dopants have been already activated. For the free carrier lifetime analysis in the case of conditions I and III, the point defects created non-conservatively during the implantation process are considered the dominant traps (N_t). A value of either 10^{15} and 10^{18}/cm^3 is used as the N_t value in the cases of low dose and high dose implantation conditions, respectively.

The free carrier lifetime was found to be a function of trap density, free carrier concentration and annealing temperature. In our analyses, a highest free carrier lifetime of about 442 ns was obtained in the case of the low dose implanted sample during the transient stage at 700°C of the dopant activation cycle (condition I). The corresponding smallest (~17 fs) free carrier lifetime was obtained for the high dose implanted sample (dopants already activated) at 1000°C, steady state part of the extended anneal cycle (condition IV). Though, in general, the free carrier lifetime was found to decrease with increasing dose, it was more so at lower temperatures (700°C) as compared to high temperatures (1000°C). Hence we suggest that any free carriers generated by irradiated photons during RTA may, at the best, influence diffusion characteristics of dopant atoms in silicon only at very low temperatures under low dose implantation conditions (Table 6).

4. Conclusions

The sheet resistance and the atomic concentration profiles of three different dopants (boron, phosphorus and arsenic), as measured by SIMS, of implanted species in silicon annealed rapid thermally in two different geometries, i.e. implanted side or wafer back side facing the heating lamps, was found to be identical, within limits of experimental error. Hence high energy infra red photons which are present exclusively during RTA, do not significantly affect both the

Table 5
Different conditions considered for our Auger analysis

Conditions	I	II	III	IV
Temperature for the Auger analysis	700°C	700°C	1000°C	1000°C
Implantation damage	taken into account	not taken into account	taken into account	not taken into account
Trap type under consideration	excess interstitials	V$^-$	excess interstitials	V$^-$
Extended defects under consideration	no	no	no	no

Table 6
Calculated Auger lifetimes for various conditions

Parameters	I		II		III		IV	
	low dose	high dose	low dose	high dose	low dose	high dose	low dose	high dose
γ (cm^6/s)	1.19×10^{-30}	1.19×10^{-30}	1.19×10^{-30}	1.19×10^{-30}	1.19×10^{-30}	1.19×10^{-30}	1.19×10^{-30}	1.19×10^{-30}
β (cm^3/s)	7.42×10^{-12}	4.31×10^{-09}	4.41×10^{-12}	8.96×10^{-08}	3.29×10^{-11}	4.33×10^{-09}	5.4×10^{-10}	6.48×10^{-10}
α (1/S)	2.26×10^6	2.26×10^{10}	4.73×10^6	1.55×10^{11}	4.17×10^8	2.54×10^{11}	1.65×10^{12}	5.68×10^{13}
Auger life time (τ, ns)	442.48	0.045	211.42	0.0065	2.40	0.004	6.1×10^{-4}	1.7×10^{-5}

dopant activation and diffusion in silicon at high temperatures. Our calculations suggest that though the free carrier lifetime increases two orders in magnitude when the annealing temperature is brought down from 1000 to 700°C, it could still be considered low enough that we may not observe any practically significant enhanced dopant activation or diffusion.

Acknowledgements

We would like to express our sincere thanks to Dr. Maggie Puga Lamberts of SIMS laboratory at the University of Florida for helping us carry out SIMS on our samples.

References

[1] Pantelides ST. In: Pantelides ST, editor. Deep centers in silicon. Gordon and Breach Science publishers, 1986. p. 1.

[2] Watkins GD. In: Pantelides ST, editor. Deep centers in silicon. Gordon and Breach Science publishers, 1986. p. 147.

[3] Peaker AR, Hamilton B. In: Pantelides ST, editor. Deep centers in silicon. Gordon and Breach Science publishers, 1986. p. 349.

[4] Lax M. Giant traps. J Phys Chem Solids 1959;8:66.

[5] Thurmond CD. The standard thermodynamic functions for the formation of electrons and holes in Ge, Si, GaAs and GaP. J Electrochem Soc 1975;122:1133.

[6] Landsberg PT. Recombination in semiconductors. Cambridge University Press, 1991.

[7] Levy RA, editor. Reduced thermal processing for ULSI. New York: Plenum Press, 1988. p. 143–80.

[8] Fair RB. Rapid thermal processing, Science and technology. Academic press, 1993. p. 349–423.

[9] Seidel TE, Lischner DJ, Pai CS, Lau SS. Temperature transients in heavily doped and undoped silicon using rapid thermal annealing. J Appl Phys 1985;57(4):1317–21.

[10] Nagabushnam RV, Singh RK, Iyer R, Sharan S, Sandhu G. Realistic three-dimensional modeling of multizone rapid thermal systems. Mater Res Soc Symp Proc 1995;387:149.

[11] Nagabushnam RV, Sharan S, Sandhu G, Rakesh VR, Singh RK, Tiwari P. Kinetics and mechanism of the C49 to C54 titanium disilicide phase transformation in nitrogen ambient. Mater Res Soc Symp Proc 1995;402.

[12] Hu H, Su L, Yang I, Antoniadis DA, Smith HI. Proc IEEE 1996 VLSI Symp Technol 1994. p. 17–8.

[13] Su L, Subbana S, Crabbé E, Agnello P, Nowak E, Schulz R, Rauch S, Ng H, Newman T, Ray A, Hargrove M, Acovic A, Snare J, Crowder S, Chen B, Sun J, Davari B. Proc IEEE 1996 VLSI Symp Technol 1996. p. 12–3.

[14] Fair RB. Modeling of dopant diffusion during rapid thermal annealing. J Vac Sci Technol A 1986;4(3):926–32.

[15] Michel AE. Anomalous transient diffusion of ion implanted dopants: a phenomenological model. Nucl Instrum Methods B 1989;37–38:379–83.

[16] Servidori M, Sourek Z, Solmi S. Some aspects of damage annealing in ion-implanted silicon: discussion in terms of dopant anomalous diffusion. J Appl Phys 1987;62(5):1723–8.

[17] Ishikawa Y, Yarnauchi K, Nakamichi I. The enhanced diffusion of low-concentration phosphorus, arsenic and boron in silicon during IR-heating. Jpn J Appl Phys 1989;28:L1319–L1321.

[18] Lojek B. Mater Res Soc Symp Proc 1991;224:33–8.

[19] Kazor A, Boyd IW. Low temperature photo-assisted oxidation of silicon. Appl Surf Sci 1992;54:460.

[20] Petroff PM. Semiconductors and insulators, vol. 5. New York: Gordon and Breach, 1983. p. 307.

[21] Troxell JR, Chatterjee AP, Watkins GD, Kimerling LC. Recombination–enhanced migration of interstitial aluminum in silicon. Phys Rev B 1979;19:5336.

[22] Van Vechten JA. Entropy of ionization and temperature variation of ionization levels of defects in semiconductors. Phys Rev B 1976;14:3539.

[23] Shockley W, Last J. Phys Rev 1957;107:392.

[24] Hirata M, Hirata M, Saito H. J Phys Soc Jpn 1969;27:405.

[25] Van Vechten JA. Activation enthalpy of recombination–enhanced vacancy migration in Si. Phys Rev B 1988;38:9913.

[26] Van Vechten JA. Simple ballistic model for vacancy migration. Phys Rev B 1975;12:1247.

[27] Fahey PM, Griffin PB, Plummer JD. Point defects and dopant diffusion in silicon. Rev Mod Phys 1989;61:289.

[28] Landsberg PT, Robbins DJ. The first 70 semiconductor Auger process. Solid State Electron 1978;21:1289.

[29] Schmid W. Experimental comparison of localized and free carrier Auger recombination in silicon. Solid State Electron 1978;21:1285.

[30] Svantesson KG, Nillson NG. Measurement of Auger recombination in silicon by laser excitation. Solid State Electron 1978;21:1603.

[31] Haug A. Carrier density dependence of Auger recombination. Solid State Electron 1978;21:1281.

PERGAMON

Materials Science in Semiconductor Processing 1 (1998) 219–230

MATERIALS
SCIENCE IN
SEMICONDUCTOR
PROCESSING

Changing from rapid thermal processing to rapid photothermal processing: what does it buy for a particular technology?

R. Singh [a,*], V. Parihar [b], S. Venkataraman [b], K.F. Poole [b], R.P.S. Thakur [c], A. Rohatgi [d]

[a]*Department of Electrical and Computer Engineering, Center for Silicon Nanoelectronics, Material Science and Engineering Program, Clemson University, Clemson, SC 29634-0915, USA*
[b]*Department of Electrical and Computer Engineering, Center for Silicon Nanoelectronics, Clemson University, Clemson, SC 29634-0915, USA*
[c]*AG Associates, San Jose, CA 95134, USA*
[d]*School of Electrical Engineering, Georgia Institute of Technology, Atlanta, GA 30322, USA*

Abstract

Rapid thermal processing (RTP) based on incoherent light as a source of energy has emerged as an alternate of furnace processing. In recent years, we have demonstrated that in addition to the high heating and cooling rates, quantum photoeffects play an important role in the operation of RTP. The quantum photoeffects can be fully exploited by using an incoherent light source as the source of optical energy (photons with wavelength less than 800 nm) and tungsten halogen lamps as the source of optical and thermal energy. The use of ultra violet (UV) or vacuum ultra violet (VUV) photons along with infrared photons is referred to as rapid photothermal processing (RPP). This paper addresses the issues involved as we move from RTP to RPP. The improved device performance results of diffusion, chemical vapor deposition (CVD), ohmic contact formation and metallization are described in detail. The manufacturing issues like stress, defects, reliability and yield are discussed in the context of RPP and the paper presents potential advantage of RPP for the manufacturing of microelectronics, solar cells, flat panel display and optoelectronics. © 1999 Elsevier Science Ltd. All rights reserved.

Keywords: Rapid photothermal processing (RPP); Solar cells; Defects; Manufacturing; Diffusion; Ohmic contacts; Microelectronics; Roughness; High *K* dielectrics; Low *K* dielectrics; Chemical vapor deposition (CVD)

1. Introduction

Since the invention of transistor in 1948, the semiconductor industry has impacted virtually every human being in one way or the other. In recent years, the microelectronics industry, due to small feature size, low-cost and high performance has enabled the introduction of low cost personal computers, leading to a revolution in information technology. Optical devices such as lasers, detectors, modulators, etc. combined with electronics have made it possible to transfer a large amount of data over the fiber network. As an energy conversion device, solar cells based on bulk Si and thin films are making a significant impact on the future sources of electrical power generation systems. Low-cost flat panel display systems are expected to play a significant role in information processing systems of the 21st century. Thus microelectronics, optoelectronics, solar cells and flat panel displays are revolutionizing the industrial world and are in fact moving towards providing new generation of similar

* Corresponding author. E-mail: singh@ces.clemson.edu.

systems in the next century. A careful examination of the above mentioned four technologies show that the past success of a particular technology depends on low-cost manufacturing and high performance of the semiconductor product. The success of microelectronics in bringing low-cost Si chip technology to a reality is mainly due to the manufacturing of high performance memory, logic and other type of semiconductor products at low-cost. With 200-mm diameter Si wafers in production and 300-mm diameter wafers to be introduced soon, the four technologies provide a scenario, where one equipment or its slight variation may be suited for all the four industries. The 300-mm diameter or larger size wafer processing system can be modified to meet the demands of all the four technologies.

Thermal processing is an integral part of semiconductor manufacturing and it covers a large number of processing steps employed in the manufacturing of any semiconductor device. The limitations of furnace processing were realized in late 1970's [1]. The research and development efforts of the last two decades have been concentrated in moving from furnace processing to rapid isothermal processing, popularly known as rapid thermal processing (RTP), which is based on incoherent light as the source of energy [2]. Due to the low thermal mass of single wafer in RTP, the heating and cooling rates are much faster than furnace processing. Reduced cycle time. a salient processing feature of RTP has provided an opportunity to solve a number of problems encountered in furnace processing [2].

As the silicon wafer diameter is approaching 300-mm, processing temperature becomes an important issue. Thermal stress as well as gravitational stress developed during the various processing steps pose limitations to the maximum temperature used in thermal processing. The maximum processing temperature for a 300-mm diameter wafer can be 900°C [3]. The issue of processing temperature is equally important for other industries. As an example, the use of low-cost glass as the substrate can lead to the development of low-cost thin film solar cells as well as low-cost flat panel display systems. Similarly, for optoelectronic devices the low processing temperature can suppress many unwanted physical phenomena associated with the fabrication of bulk optical devices. The low processing temperature can also allow the monolithic integration of various optical devices on the common substrate. Thus the issue of low processing temperature is a common problem being faced by the IC industry, the solar cell and flat panel display industry as well as the optoelectronics industry. Defects affect performance, reliability and yield drastically. For the realization of advanced silicon ICs based on 180 nm and lower feature sizes, the defect reduction is more important than it has been in the past. Therefore as we are moving towards the 21st century, stringent defect

reduction schemes need to be developed [4]. These stringent defect reduction and yield enhancement issues in manufacturing are difficult to meet with RTP. Based on our fundamental contribution about the role of photoeffects in RTP [5–10], we have exploited the quantum photoeffects in developing rapid photothermal processing (RPP) in place of RTP. In case of rapid photothermal processing (RPP), ultra violet (UV) or vacuum ultra violet (VUV) lamps are employed in addition to the tungsten halogen lamps used in RTP. This paper discusses the evolution of RPP from RTP and the potential advantages it has to offer.

2. Background material

The equation,

photons + matter = thermal effects + quantum effects

describe the photon–matter interaction. Fig. 1 shows the spectral contents of various incoherent light sources. Thermal effects dominate for photons with wavelength greater than about 800 nm. Photons in this range increase the vibration amplitude of the atoms, raising the temperature of the film and the underlying substrate so that the desired thermal effect takes place. Photons of wavelength less then 800 nm excite the atoms and molecules participating in thermal process electronically and quantum effects are observed. As shown in Fig. 2, for light sources with wavelength in between 400–800 nm, thermal effects as well as quantum effects are observed. On the other hand, for photons with wavelength between about 100–400 nm quantum effects are observed [11]. Fig. 2 shows the difference between a purely thermal process and a photothermal process [10]. In case of photothermal process, the quantum potential barrier is significantly reduced. As a result, the processing temperature, processing time as well as microscopic defects are lower in a photothermal process than the corresponding pure thermal process. The main features of RPP are described below (Section 2.1).

2.1. Reduced activation energy

As shown in Figs. 1 and 2, the incident photon with wavelength less than 800 nm reduces the potential barrier and hence the activation energy required for diffusion is decreased and this results in an enhancement of the diffusion coefficient. In a recent publication [12], we have shown that the (diffusion coefficient)$_{photo-thermal}$ > (diffusion coefficient)$_{thermal}$. The diffusion coefficient is given by

$$D = fn_c\lambda^2 v_0 \exp((\Delta S_f + \Delta S_m)/k)\exp((-\Delta H_m + \Delta H_f)/kT),$$

Photons + Matter = Thermal Effects + Quantum Effects

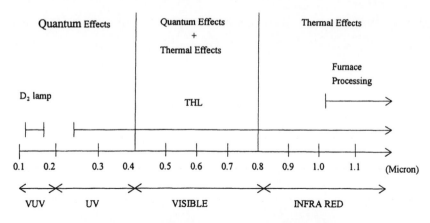

Fig. 1. Spectral dependence of thermal and quantum effects.

where λ is the interatomic distance, v_o is the atomic vibrational frequency of solid, ΔS_f and ΔS_m are the entropies of formation and migration and ΔH_f and ΔH_m are the enthalpies of formation and migration, respectively. The correlation factor f is 0.78 and co-ordination factor n_c is 4 for silicon. The energy required for the formation of vacancies is very high and therefore ΔS_f and ΔH_f remain unchanged for a given processing temperature. Therefore the net effect is an increase in the bulk diffusion coefficient. Hence, for excited electronic states, the atomic diffusion occurs at a lower temperature.

The reduction of activation energy in RPP has direct effect on the reduction of activation energy of mass transport limited region in the chemical vapor deposition (CVD) system. In a recent publication [13], we have shown the importance of high-energy photons ($\lambda < 800$ nm) in reducing the activation energy of mass transport limited region and providing higher growth rates and low defect densities for the material deposited by RPP assisted CVD. This observation has

direct advantage of depositing various electronic and optical materials by CVD at low processing temperatures.

2.2. Reduced surface roughness and reduced interface roughness

The surface roughness as well as related interface roughness plays a very important role in the operation of semiconductor devices. Statistical roughening is primarily responsible for the formation of surface roughness and defects. The statistical fluctuation of the incoming atoms is reduced with lower temperature and this results in higher uniformity. Statistical roughening arises because of statistical fluctuation in the arrival rate of vapor flux. The fluctuation in flux increases with the increase of the processing temperature. During the deposition surface diffusion and statistical roughening compete with each other, the first increasing the film roughness and the second smoothing it out. In order to decrease the roughness, we can either

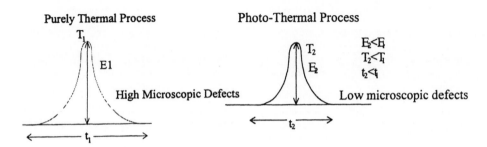

Fig. 2. Comparison between purely thermal process and photo process in terms of the activation energy, microscopic defects, processing temperature and cycle time.

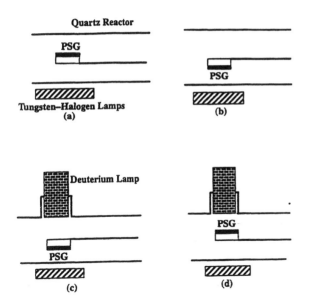

Fig. 3. Different lamp configurations.

Table 2
Diffusion at 700°C with RPP

Experimental setup	Sheet resistance (Ω/sq)	Minority carrier lifetime in bulk (injection level of 1×10^{15} carriers/cm^3) (μs)
Back irradiation Fig. 3(a)	no diffusion	8.13
Front irradiation Fig. 3(b)	no diffusion	16.2
Back irradiation + VUV (front) Fig. 3(d)	71	341.1

increase the surface diffusion or decrease the statistical roughening. In RTP, the surface diffusion coefficient is low and strongly dependent on temperature. Thus, increasing the surface diffusion becomes the dominant factor in decreasing the roughness. We mentioned earlier that the bulk diffusion coefficient is higher for RPP. Similarly, in RPP, the surface diffusion coefficient is higher at a given temperature. Thus the use of a low processing temperature and higher diffusion coefficient in RPP leads to lower surface roughness [14].

2.3. Reduced stress and lower microscopic defects

Since the bulk and surface diffusion coefficients are higher in RPP, the overall energy of the system is reduced. This is due to the photoexcitation of the vapor flux, the substrate as well as any over layer. The lower energy of the system leads to the reduction of intrinsic stress as well as microscopic defects.

2.4. Lower cycle time

With reduced potential barrier and higher diffusion coefficient we can carry out any process step with a reduced thermal cycle. As a result the throughput of the process is increased.

3. Experimental results

In this section we present experimental results of diffusion, ohmic contact formation and metallization with RPP. In this section we also present the RPP assisted CVD results of high and low dielectric constant materials.

As a source of VUV photons, we have used a deuterium lamp in our RPP system in conjunction with tungsten halogen lamps. As shown in Fig. 3, we have used various lamp configurations to demonstrate the role of high-energy photons in RPP. In case of ohmic contact formation we have used an UV lamp in conjunction with a tungsten halogen lamp.

Table 1
Results for various lamp configurations

Diffusion Configuration	Sheet resistance (Ω/sq)	Junction depth (X_j) (μm)	Surface concentration (atoms/cm^3)	Minority carrier lifetime (μs)	Leakage current density (A/cm^2)
Back THL irradiation Fig. 3(a)	16.98	0.50	2.3×10^{23}	1.30	1×10^{-10}
Front THL radiation Fig. 3(b)	10.19	0.58	4.0×10^{20}	6.20	7×10^{-11}
Front THL + back VUV irradiation Fig. 3(c)	10.18	0.62	4.1×10^{20}	8.81	3×10^{-11}
Back THL + VUV irradiation Fig. 3(d)	10.04	0.69	5.0×10^{20}	9.85	1×10^{-11}

Table 3
Back contact formation with RPP

Case	Lamp configuration	XRD ratio of peak intesities of (111) and (200)	Stress (dynes/sq·cm)
A	FI + back UV	3.04	2.334×10^8
B	BI + front UV	2.5	8.66×10^9
C	BI	2.3	1.28×10^{10}
D	FI	1.853	1.433×10^{11}
E	FI + increased time	2.1	5.03×10^{10}
F	FI + increased temperature	2.152	8.44×10^{10}

3.1. Diffusion

A spin-on phosphosilicate glass (PSG) with a concentration of 10^{21} cm^{-3} was used as n-diffusion source. P-type silicon wafers were heated from room temperature to 880°C at a rate of 30°C/s and diffusion was carried out for 30 s. The samples were cooled at a rate of 0.667°C/s until 800°C and then rapidly cooled to room temperature in 3 min. Table 1 [15] shows the results of the diffusion study. As shown in Table 1, case (d) has provided the best results, the difference among the different cases being the availability of photons of different wavelength at the diffusion front. In the case of back irradiation, only photons of wavelength greater than 800 nm are reaching at the surface, since the wafer is 400 µm thick. In the case of front irradiation, the visible and UV photons with wavelength between 300–800 nm are directly incident on the surface. We have observed a one-to-one correlation between the spectrum of photons available at the diffusion front and the diffusion coefficient [15].

We were able to carry out diffusion at 700°C with the RPP system of Fig. 3(d). No diffusion was observed in the absence of the high-energy VUV photons. The results are shown in Table 2 [12]. The higher activation energy and an increase in solid solubility of the dopant are the main reasons for higher surface concentrations observed with high-energy photon assisted diffusion. The gettering mechanism is more effective in the presence of UV and VUV photons due to photo induced electronic excitations of the dopant and the molecules of the silicon substrate. This explains the higher bulk minority carrier lifetime value obtained for VUV irradiated samples [15].

3.2. Ohmic contacts formation and metallization

The role of photoeffects in the formation of aluminum back surface (BSF) contacts on p-type wafer has been studied using different lamp configurations. Lamp arrangements similar to Fig. 3 were used in this study and the results are shown in Table 3. The results clearly demonstrate that the irradiation of UV photons on the BSF paste produces the best results. These results are better than those with increased process time and without the use of UV photons [16]. The RPP processed samples show better uniformity at the interface and lower stress. Thus, RPP provides microstructure homogeneity.

Alloys of Al–Y (wt% of 0.1) were sputtered on oxidized silicon wafers. The sputtered films were annealed using furnace annealing, rapid thermal annealing and rapid photothermal annealing. The resulting electrical resistivity data and stress values are shown in Table 4 [17]. Furnace annealing was done at 425°C for 30 min. It is interesting to note that the same resistivity values were obtained by rapid thermal annealing and

Table 4
Comparison of electrical resistivity and stress values for Al–Y samples annealed with different thermal processing techniques

	Resistivity (µΩ cm)	Stress As-deposited (compressive) (dynes/cm^2)	Stress annealed (tensile) (dynes/cm^2)
Furnace annealing (425°C for 30 min)	2.71	3.2×10^8	2.13×10^8
Rapid thermal annealing (400°C for 5 min)	2.67	3.36×10^8	1.05×10^8
Rapid photothermal annealing (350°C for 5 min)	2.67	3.29×10^8	8.94×10^7

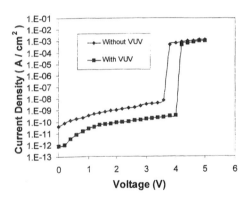

Fig. 4. J–V characteristics of Ta_2O_5 films annealed at 600°C with and without VUV photons.

rapid photothermal annealing for 5 min at 400 and 350°C, respectively.

3.3. Rapid photothermal chemical vapor deposition (RPCVD)

The RPP assisted CVD system is designed in such a way that UV and VUV photons are incident directly on the side of the substrate where chemical reactions take place. For the first time, we used RPCVD for the deposition of high temperature superconductors [18]. Since then, we have used RPCVD for the deposition of a number of material systems. Here, we have focussed on the deposition and characterization of high dielectric constant and low dielectric constant materials.

3.3.1. High K dielectrics

For the manufacturing of next generation of ICs, there is a need for developing high dielectric constant materials for dynamic random access memory (DRAM) capacitors as well as for gate dielectric appli-

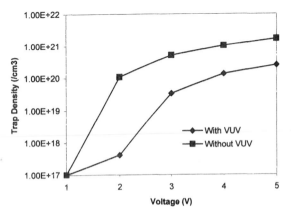

Fig. 5. Trap density distribution for the Ta_2O_5 sample annealed by RTA and RPA.

cations. Leakage current is a significant problem and, as described in the following paragraph, we have produced better results with RPP than any other deposition technique.

Ta_2O_5 films were deposited by thermal CVD and annealed in the RPCVD system Fig. 4 [11] shows the J–V characteristics with and without the VUV photons. Trap density distribution for the Ta_2O_5 films annealed with and without VUV photons are shown in Fig. 5 [11]. The results of Fig. 5 show that in addition to the improved performance, RPP also provides better reliability than RTP. We also deposited lead lanthanum zirconia titanate (PLZT) films and observed lowest leakage current density value ever reported for any non-Si-based inorganic dielectric materials [11]. In case of annealing of PLZT films we obtained significantly lower surface roughness and microcracks [11]. The low leakage current as well as lower trap generation during voltage stress results show the importance of RPP over RTP in obtaining better performance and reliability of semiconductor devices.

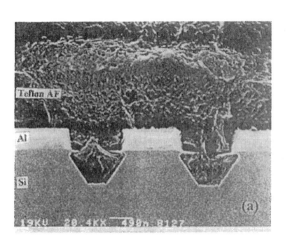

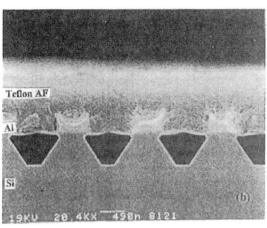

Fig. 6. Cross sectional SEM of the Teflon AF® gap fill obtained for sample processed (a) with the use of VUV photons and (b) without the use of UV photons.

3.3.2. Low K dielectrics

These materials are important as dielectrics for interconnect in ICs and multichip modules (MCMs) as they reduce power consumption, crosstalk and signal propagation delay. Teflon ($K \sim 1.9$) was deposited using a direct injection and RPP assisted CVD system. Fig. 6 [11] shows the cross sectional SEM of the sample deposited without using VUV photons and Fig. 6(b) shows the same for a sample processed using VUV photons. Fig. 6(a) indicates that there are voids in the gaps. The surface diffusion coefficient is low and as a result the Teflon particles coalesce with each other. This does not permit the reacting spices to move about into the gaps and thus they form a bridge over the gaps. Fig. 6(b) shows a clear improvement in gap-fill of Teflon AF film by using VUV photons. Gaps are completely filled and do not show any voids. In the presence of VUV light, Teflon AF particles have enough surface migrational energy to diffuse along the surface and fill the gaps. These results demonstrate that the use of VUV photons leads to the realization of conformal coatings at low processing temperature. Thus, RPP is a superior back end processing technique.

3.4. Solar cell

As early as 1985 [19], we demonstrated the importance of rapid photothermal processing in the fabrication of solar cells. A complete list of various advantages offered by RPP is given in Ref. [2]. In a 1986 publication [5], we demonstrated the importance

Table 5
RTP-diffused silicon solar cells fabricated at Georgia Tech. (All cells tested and verified by Sandia National Laboratories, except for[a])

Materials	Resistivity (Ω cm)	V_{oc} (mV)	J_{sc} (mA/cm^2)	FF	Eff (%)	Process	Area (cm^2)	Year
FZ	1.3	632	37.3	0.808	19.1	RTP/10 µm-Al–BSF/RTO/PL	4	1997
FZ	2.3	627	37.4	0.801	18.8	RTP/SP-Al–BSF/RTO/PL	4	1997
FZ	1.3	624	36.4	0.812	18.5	RTP/5 µm-Al–BSF/RTO/PL	4	1997
FZ	2.3	616	37.2	0.794	18.2	RTP/5 µm-Al–BSF/RTO/PL	4	1997
FZ	2.3	612	36.0	0.773	17.1	RTP/5 µm-Al–BSF/RTO/PL	42	1997
FZ	2.3	608	35.9	0.766	16.7[a]	IR-Belt/PECVD/SP	4	1997
FZ	1.3	628	33.9	0.798	17.0	RTP/SP-Al-BSF/RTO/PECVD/SP	4	1997
FZ	1.3	623	36.9	0.809	18.6[a]	RTP/RTO/PL	1	1996
FZ	0.6	633	35.9	0.789	17.9	RTP/RTO/PL	4	1996
FZ	2.2	612	34.0	0.785	16.3	RTP/PECVD/SP	4	1996
FZ	0.2	633	31.4	0.773	15.4	RTP/PECVD/SP	25	1996
FZ	0.2	637	32.8	0.819	17.1	RTP/PL	1	1994
FZ	0.2	627	34.9	0.779	17.0	RTP/PL	1	1994
Bayer Cz	1.3	613	37.1	0.811	18.4	textured RTP/5 µm-Al–BSF/RTO/PL	4	1997
Wacker Cz	2.8	607	37.5	0.790	18.0	RTP/RTO/PL	1	1996
Wacker Cz	2.8	605	33.9	0.774	15.9	RTP/PECVD/SP	4	1995
Wacker Cz	2.8	609	35.2	0.763	16.4	RTP/PL	1	1994
Solarex mc-Si	1–2	618	34.0	0.795	16.7[a]	RTP/5 µm-Al–BSF/RTO/PL	1	1997
Solarex mc-Si	1.0	598	30.4	0.762	13.8	RTP/SP-Al–BSF/RTO/SP	4	1997
HEM mc-Si	1.0	610	34.8	0.768	16.3[a]	RTP/5 µm-Al–BSF/PL	4	1997
HEM mc-Si	0.9	593	34.1	0.790	16.0	RTP/SP-Al–BSF/RTO/PL	1	1996
OTC mc-Si	0.8	594	33.0	0.756	14.8	RTP/PL	1	1994
EFG mc-Si	2–4	581	36.0	0.767	16.0[a]	RTP/5 µm-Al–BSF/RTO/PL	1	1997
EFG mc-Si	2–4	569	31.6	0.795	14.3	RTP/PL	1	1995
Dendritic Web	11 (n-base)	618	35.4	0.792	17.3	RTP/CFO	1	1994
Dendritic Web	11	559	33.9	0.783	15.1	RTP/PL	1	1994
Silicon Film℗	1–2	545	31.3	0.677	11.6	RTP/PL	1	1995

[a] Measured at Georgia Tech.

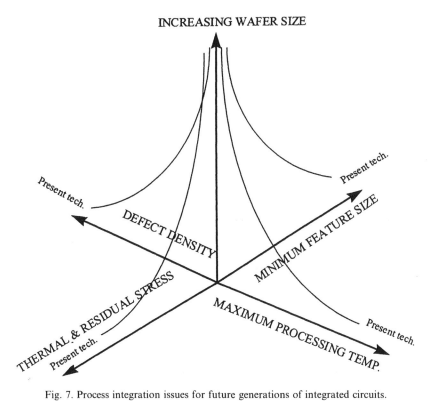

Fig. 7. Process integration issues for future generations of integrated circuits.

of RPP in reducing the processing time of solar cells. Since 1985, we have published experimental results on the various processing steps used in the fabrication of solar cells [2, 6–8, 12, 15, 20–22]. As a result of our fundamental contributions, researchers are implementing RPP for the manufacturing of solar cells [23]. Researchers at Georgia Tech University have fabricated high efficiency bulk silicon solar cells by RTP [24]. A summary of the experimental results of the authors of Ref. [24] is given in Table 5.

4. Process integration, defects and manufacturing issues in microelectronics

With the feature size decreasing, manufacturing in the 21st century will be driven by a number of challenges. New materials, defect density, defect distribution, uniformity of the material, magnitude and distribution of thermal stress and process optimization become significant issues as the industry enters into the nanoelectronics domain. Defects will be the major issue in determining the performance, yield and reliability of a semiconductor product. Fig. 7 [17] shows the process integration issue for future generation ICs. This figure clearly shows that the future generation of semiconductor processing will require reduced defect density and lower process temperature. The introduc-

tion of recombination centers occurs at higher processing temperatures, thereby reducing minority carrier lifetime. As has been demonstrated in Section 3, with RPP we can reduce the processing temperature thereby increasing the carrier lifetime. It has been observed from our experimental results that RPP gives reduced defects and stress in both front and back end processing. This clearly indicates the usefulness of RPP as an efficient process integration technique.

High surface migration, high uniformity and low processing temperature are essential to lower the number of defects [11]. The optical energy reduces the activation energy and increases surface migration. As mentioned before, statistical fluctuations are reduced

Table 6
Advantages of RPP over RTP for the manufacturing of microelectronics

Parameter	RTP	RPP
Processing temperature	high	low
Throughput	low	high
Stress	high	low
Defects	high	low
Surface roughness	high	low
Reliability	low	high
Process integration	difficult	easy

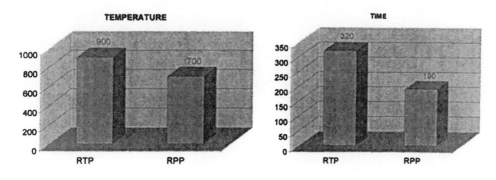

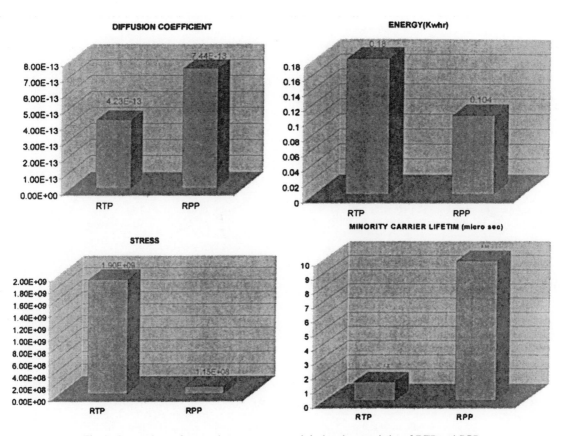

Fig. 8. Comparison of processing parameters and device characteristics of RTP and RPP.

with reduction in processing temperature, thereby reducing surface roughness. The thermal stress developed during various processing steps is also reduced with processing at lower temperature. The improved minority carrier lifetime demonstrates that there is a decrease in microscopic defects. Thus RPP paves the way for achieving a homogenous microstructure which is very essential as we move into the nanoelectronics era. Thus, changing from RTP to RPP will provide a lower processing temperature, resulting in higher yield,

performance and reliability. Table 6 compares the RTP and RPP techniques and Fig. 8 compares RTP and RPP in the context of processing parameters and device performance. In case of diffusion, similar results have been reported by other researchers [25].

Fig. 9 shows the past, present and future trends of thermal processing, lithography and defect detection techniques. The lithography and defect detection tools are going to employ lower wavelength ions, electrons and photons with lower wavelength. Thus, the use of

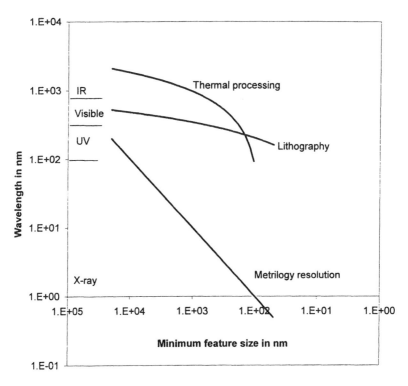

Fig. 9. Trends in semiconductor industry: moving towards smaller wavelengths.

high-energy photons in thermal processing is in concert with the trends in lithography and defect detection techniques.

5. Role of RPP in the manufacturing of solar cells, flat panel display and optoelectronics

As we mentioned earlier, the key features of RPP can lead to a breakthrough in the manufacturing of bulk and thin film solar cells as well as flat panel display manufacturing. In the case of solar cells, a demonstration of high efficiency is not sufficient without addressing the manufacturing issue. Thin film technology like $CuInSe_2$, CdTe, etc. cannot penetrate the market due to the lack of defect reduction and yield issues. Hence defect reduction and yield issues are very important in the manufacturing of solar cells. The key features of RPP and their direct effect on manufacturing of a particular technology is listed in Table 7. RTP has not been widely used in the manufacturing of optoelectronics. However based on the results

Table 7
Advantages of RPP for solar cells and flat panel display manufacturing

Salient features of RPP	Si bulk solar cell manufacturing	Thin film solar cell manufacturing	Flat panel display manufacturing
Low processing temperature	improved transport properties leading to high efficiency	low cost substrate and high efficiency	low cost substrate and high performance transistors
Short processing time	high throughput	high throughput and unwanted chemical reaction between substrate and cells can be avoided	high throughput and no degradation of the transistor properties
Low defect and process uniformity	high efficiency	yield of thin film cells is high	yield of thin film transistors is high

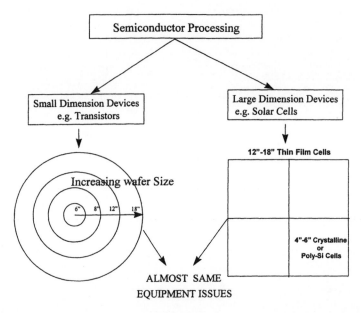

Fig. 10. Equipment issues for microelectronics, optoelectronics, solar cells and flat panel displays manufacturing.

presented for other technologies, we expect a similar beneficial effect of RPP in the manufacturing of optoelectronics.

6. Equipment related issues

As shown in Fig. 10, RPP based equipment provide a common solution for all the four industries, i.e. microelectronics, optoelectronics, solar cells and flat panel displays. Due to small device dimensions, the microelectronics industry requires a much tighter tolerance control on processing conditions, whereas, due to large device size of solar cells the requirements will be moderate. The other two industries' requirements fit in between these two extreme cases. The new approach in RPP equipment design, can provide a RPP-based manufacturing technique to cover a wide range of processes. In one extreme case, RPP can be used in continuous batch processing. The other extreme end is single batch processing. The research and development cost of new RPP equipment will be lower, since modified design changes (e.g. different control systems) can provide a larger market share for almost the same investment of money.

7. Conclusion

For the manufacturing of microelectronics, RPP has been identified as a better processing technique for both front end and back end processing. It has also been shown that it combats the common technology challenges being posed in the manufacturing of optoelectronics, solar cells and flat panel displays. Results have been presented which reiterate our claim that RPP is a lower temperature, lower stress and lower defects processing technique. Manufacturing issues have been considered and RPP has been proposed as the thermal processing technique to meet future challenges in the semiconductor industry.

References

[1] Ferris SD, Leamy HJ, Poate JM, editors. Laser solid interactions and laser processing, 1978. In: AIP Conf. Proc. No. 50. New York: American Institute of Physics, 1979.

[2] Singh R. J Appl Phys 1988;63:R69–R114.

[3] Huff HR, Goodall RK, Nilson RH, Griffiths SK. 191st Electrochem Soc Meet ECS Ext Abstr Vol 1997;97-1:745.

[4] R Singh, KF Poole. In: KLA-Tencor Yield Management Solutions Seminar, Section 6, 1998. p. 2–17.

[5] Singh R. Semiconduct Int 1986;9(1):28.

[6] Singh R, Radpour F, Chou P. J Vac Sci Technol A 1989;7:1456.

[7] Singh R, Chou P, Radpour F, Nelson AJ, Ullal HS. J Appl Phys 1989;66:2381.

[8] Singh R, Sinha S, Thakur RPS, Chou P. Appl Phys Lett 1991;58:1217.

[9] R Singh. Rapid thermal processing. In: Holloway P, McGuire G, editors. Handbook of compound

semiconductors. Park Ridge, NJ: Noyce Publications, 1995. p. 442–517.

[10] Singh R, Sharangpani R. Solid-State Technol 1997;40(10):193.

[11] Singh R, Nimmagadda S, Parihar V, Chen Y, Poole KF. IEEE Trans. Electron Device 1998;45:643.

[12] Singh R, Cherukuri KC, Vedula L, Rohatgi A, Narayanan S. Appl Phys Lett 1997;70:1700.

[13] Singh R, Vedula L, Gong C. J. Electron Mat 1998;27:L13.

[14] Singh R, Chen Y. J. Electron Mat 1997;28:1184.

[15] Singh R, Cherukuri KC, Vedula L, Rohatgi A, Mejia J, Narayan S. J. Electron Mat 1997;26:1422.

[16] Singh R, Vedagarbha V, Nimmagadda SV, Narayan S. J Vac Sci Tech B 1998;16:613.

[17] R Singh, S Nimmagadda, V Parihar, Y Chen, KF Poole, L Vedula. IEEE Trans Semiconduct Manuf, in press.

[18] Singh R, Sinha S, Hsu NJ, Ng JTC, Chou P, Narayan J, Singh RK. J Appl Phys 1991;69:2418.

[19] Singh R. Electronics 1985;58:19.

[20] Singh R, Radpour F, Nguyen Q, Chou P, Joshi SP, Ullal HS, Mattoon RJ, Asher S, Kazmerski LL. J Vac Sci Tech A 1987;5:1819.

[21] Thakur RPS, Singh R. Appl Phys Lett 1994;64:327.

[22] Mavoori J, Singh R, Narayanan S, Chaudhuri J. Appl Phys Lett 1994;65:657.

[23] Szluficik J, Sivoththaman S, Nijsr JF, Mertens P, Overstraeten RV. Proc IEEE 1997;85:711.

[24] Rohatgi A, Doshi P, Kamra S. 14th Eur Photovoltaic Spec Conf Proc 1997;1:660.

[25] Noel S, Ventura L, Slaoui A, Muller JC, Groh B, Schindler R, Froschle B, Theiler T. Appl Phys Lett 1998;72:2583.

PERGAMON

MATERIALS
SCIENCE IN
SEMICONDUCTOR
PROCESSING

Materials Science in Semiconductor Processing 1 (1998) 231–236

Phosphorus diffusion from a spin-on doped glass (SOD) source during rapid thermal annealing

D. Mathiot[a],*, A. Lachiq[a], A. Slaoui[a], S. Noël[a], J.C. Muller[a], C. Dubois[b]

[a]*PHASE-CNRS, 23 Rue du Loess, B.P. 20, F-67037 Strasbourg, France*
[b]*LPM, INSA de Lyon, 20 Av. A. Einstein, F-69621 Villeurbanne Cedex, France*

Abstract

Limiting thermal exposure time using rapid thermal processing (RTP) has emerged as a promising simplified process for microelectronics applications and for manufacturing of terrestrial solar cells in a continuous way. Especially, rapid thermal diffusion (RTD) of phosphorus from doped oxide films (SOD) was extensively used for the emitter formation purpose but few work concerned the diffusion mechanism.

Here we investigate more in details the diffusion kinetics of phosphorus after rapid thermal annealing of P-SOD coated silicon samples. The observed enhanced distribution of phosphorus after RTD is discussed based on the dopant sources and processing conditions. Comparisons between experimental profiles and simulation results using up to date phosphorus diffusion models allow us to discriminate between various possible enhancement mechanisms. © 1999 Elsevier Science Ltd. All rights reserved.

1. Introduction

Shallow junctions formation in silicon is one of the most important steps in the fabrication of microelectronics (IC's) and photovoltaic (PV) devices. Because of various problems associated with ion implantation and subsequent annealing [1, 2], thermal diffusion of dopants into the silicon substrate from a solid, liquid or gaseous source has been widely investigated [3–5]. Usually, diffusion is accomplished by heat treatment in a classical furnace at temperatures above 900°C for times ranging from minutes to hours. However, conventional furnace heating for prolonged periods of time results in deep junctions which are unsuitable for common applications. Recently, rapid thermal diffusion (RTD) using various dopant sources such as planar dopants or doped-spin-on glass (SOD) films deposited on processed wafers, were reported in the literature [6–10]. First, coating with glass films is a very simple process and causes no damage. Next, the

association of SOD films with rapid thermal annealing (RTA) using radiation from tungsten halogen lamps as a heat source offers many attractive features such as low cost, minimum overall thermal budget, low power consumption and high throughput. The characteristic features of a RTA temperature profile are a high heating rate, then a plateau at a temperature T with time duration from a few seconds to a few minutes, followed by a rapid cool-down which will inherently include a 'quenching' effect. The total time taken to reach room temperature is less than 1 min. This makes this process better adapted to µm ICs and for controlled emitter for PV cells than other methods.

Doping characteristics from phosphorus doped oxide glass (P-SOD) for conventional furnace diffusion have been reported by many researchers [11–14]. So far however, insufficient published data for RTP doping are available to clarify doping mechanism [10, 13].

In this paper, experimental results on RTP doping from P-SOD are described. Sheet resistance and phosphorus depth profiles for diffused layers are measured as a function of diffusion time, diffusion temperature and phosphorus concentration in the P-SOD source. Moreover, systematic comparisons with simulated P

---

* Corresponding author. Tel.: +33-3-8865-5102; fax: +33-3-8865-5179; e-mail: mathio@erm1.u-strasbg.fr

Table 1
Phosphorus concentration in the SOD solutions

solutions	P505	P506	P507	P509
P concentration ($\times 10^{20}$ cm^{-3})	0.45	3	5	20

profiles calculated with an up-to-date model, taking into account all the known complex couplings between the diffusing dopant and the silicon point defects [15] permit to make some suggestions on the possible origin of the observed enhanced diffusion.

2. Experiment

The silicon samples used in the experiments were 450 μm thick, ⟨100⟩ oriented, 1–10 Ω·cm, p-type (boron doped) crystals. They were 15×15 mm^2 in size, cut from standard one-side polished wafers. The phosphorus doped SOD solutions used here were commercially available from Filmtronics Co. (USA) and contains a high phosphorus dopant concentration. The dopant concentration for the doped solutions are listed in Table 1. These solutions contain ethyl alcohol as a solvent and may be diluted with methanol or isopropanol to modify viscosity and obtain controlled oxide film thickness. For the deposition of the SOD films, the spin-on technique was used. Various film thicknesses were achieved by diluting the solutions with methanol. The wafers were spun at a speed of 9900 rpm/min for 15 s, followed by a prebaking at 200°C for 15 min using a hot plate before placing them into the RTP furnace.

Rapid thermal processing was carried out at 800–950°C for 2–120 s in Ar ambient in a commercial tungsten halogen lamp system (FAV4 JIPELEC, Grenoble). For comparison purpose, some P-SOD coated samples were annealed in an open-tube furnace (classical thermal processing) at similar temperatures for 15 min.

Measurements of sheet resistance (R_s) were performed on the processed wafers by using the four-point probe method after stripping the SOD layer in a HF solution. Phosphorus depth profiles in Si were measured using secondary ion mass spectroscopy (SIMS).

3. Results

Fig. 1 reports the evolution of the sheet resistance R_s as a function of temperature after rapid thermal diffusion of phosphorus from the SOD films. The sheet resistance measurements give an estimation of the dopant element activation level. R_s decreases sharply with an increase in the processing temperature for all solutions. This behavior is caused by the diffusion and subsequent activation of phosphorus atoms into silicon. Thus, RTP seems to be very efficient for diffusion from a SOD layer. Sheet resistance values lower than 150 Ω/sq are easily obtained for temperatures higher than 850°C. In particular, at temperatures generally used for classical dopant diffusion processes (i.e. 850–950°C), we reached the typical sheet resistance range 40–100 Ω/sq suited for many applications after only 25 s in a lamp furnace, instead of 15 min in a conventional furnace. This observation is a strong indication that P diffusion is significantly enhanced for this experimental condition, as compared with the conventional furnace anneal condition.

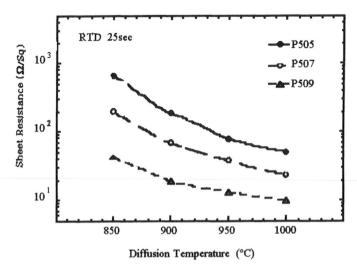

Fig. 1. Sheet resistance versus RTD temperature for P-SOD coated samples using different doping solutions.

As can also be seen in Fig. 1, the higher is the phosphorus concentration in the solution, the lower is the sheet resistance. This result confirms that doping is limited by the supply of phosphorus from the SOD layer source. The doping efficiency dependence on the SOD source has also been found in RTD [9, 10] and furnace [16] experiments. For example, thinner oxide layers resulted in higher sheet resistance values, which in turn indicates smaller dopant release from the oxide and total dopant (Q) in oxide indicative of finite dopant sources.

A typical experimental SIMS phosphorus profile is shown in Fig. 2 (curve a). This particular example corresponds to a RTD at 850°C for 90 s, using the undiluted P506 SOD source. Also shown in this figure is the theoretical profile simulated using the ATHENA process simulator. For the calculation we used the so-called 'CNET' model which uses the theoretical approach described in [15] and has been proven to provide very accurate results for high concentration furnace P diffusion [17]. The comparison of the two curves clearly confirms the enhanced behavior caused by the SOD-RTD process.

4. Origin of the diffusion enhancement

In this section we will report on some specific experiments and simulations performed to try to elucidate the physical origin of the diffusion enhancement.

4.1. RTP versus furnace diffusion

We first checked if this anomalous behavior was linked to the specific kind of diffusion source (SOD) used in this study. For that we performed some conventional furnace diffusion runs using the same diffusion source. A typical result is given in Fig. 3 for the 900°C, 15 min case (curve a). Also shown in the figure is the simulated profile (curve b), using the model mentioned above, without any parameter fitting (i.e. using the usual default parameters of the model). It is obvious that the agreement between the calculated and the experimental profiles is excellent. This is a strong indication that, when performed in conventional furnace condition, phosphorus diffusion from a SOD source occurs normally. We are thus led to the conclusion that RTP is required to observe the enhancement.

4.2. Surface concentration

An intrinsic property of the RTP treatments is the fast temperature ramping-up and the short high temperature dwell time. As a consequence any transient phenomena, associated with temperature and/or interface chemical equilibration, can have a higher importance than for conventional furnace anneal.

Fig. 4 reports the measured P surface concentration in the Si substrate as a function of the RTD time at 850°C. This concentration is extracted from the SIMS profiles measured after stripping off the SOD source

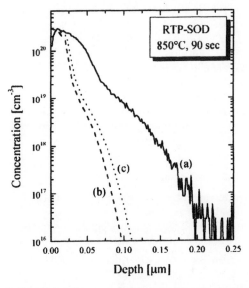

Fig. 2. Comparison between experimental (a) and calculated P diffusion profiles at 850°C for 90 s (RTP–SOD). Calculation of (b) assumes a constant surface concentration, whereas curve (c) was obtained by taking into account the initial transient surface concentration shown in Fig. 4.

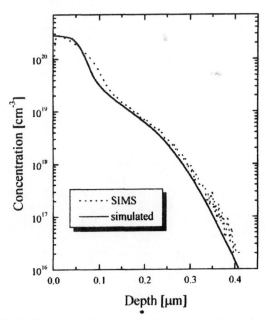

Fig. 3. Comparison between experimental (a) and calculated (b) P diffusion profiles at 900°C for 15 min (furnace anneal).

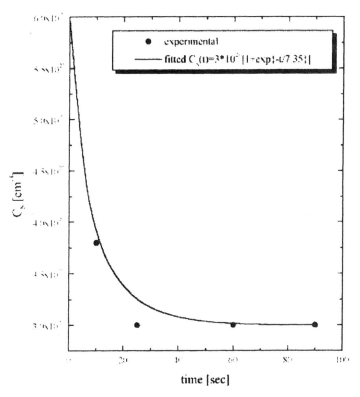

Fig. 4. Variation of the phosphorus concentration at the Si surface, as a function of the anneal duration of the RTP-SOD treatment.

film. From this figure, it is clear that the surface concentration starts from a rather high value and then decreases up to its asymptotic equilibrium value, which is reached after only a few tens of seconds. Since the P diffusion strongly depends on the surface concentration, this behavior is expected to contribute to the observed enhanced diffusion. To quantify this point, a simulation was performed by taking into account this surface concentration variation according to the decaying exponential law

$$C_{surf} = 3 \times 10^{20}[1 + \exp(-t/\tau)](cm^{-3}),$$

with $\tau = 7.35$ s. The results of this simulation is given as curve (c) in Fig. 2. Although the calculated profile is significantly deeper than curve (b) (calculated with a constant surface concentration equal to the asymptotic value), the experimental profile is still largely deeper. Thus, this transient surface concentration behavior is not enough to account for the observed enhancement.

4.3. Influence of the SOG film

In order to gain further insights on the respective influence of the RTP and the presence of the SOD diffusion source we performed the following experiments:

A first diffusion RTP anneal was performed at 900°C for 60 s, using a diluted P506 SOD source. The corresponding SIMS profile is given in Fig. 5 (curve a). Then the highly doped SOD film was removed by HF etch and subsequent anneals were performed in various conditions:

1. RTP at 900°C/60 s in Ar ambient, with the bare Si surface (curve b in Fig. 5).
2. deposition of undoped spin on glass, and then RTP in the same condition as previously (curve c in Fig. 5).

It is found that the resulting final profiles are very closed to each other, the small differences being within the SIMS accuracy limit.

Curve (d) of Fig. 5 is the calculated profile using the 'CNET' model of the ATHENA simulator, starting from the measured curve (a) as the initial profile. As can be seen, the agreement between curves (b), (c) and (d) is excellent. This proves that the diffusion occurs normally during the second anneal, independently of the presence or not of the undoped SOG film. This indicates that the diffusion enhancement observed during RTP-SOD treatments is not due to the RTP itself, nor

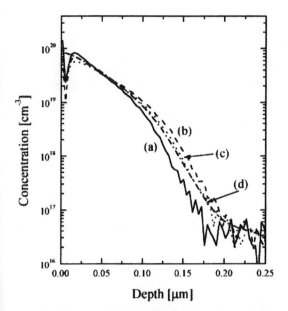

Fig. 5. Phosphorus profiles after redistribution of a RTP-SOD diffused initial profile (a) under various conditions. (b) RTP in inert ambient, bare sample. (c) RTP in inert ambient after undoped SOG coating. (d) Simulated profile corresponding to case (b).

to the presence of the surface film itself (because of interfacial stress effects for example).

It is further emphasized that the use of a diffusion model including the coupling with the Si point defects (pair diffusion model) is required to obtain a good fit of this experiment. Indeed, the 'FERMI' model of ATHENA (which takes into account only the Fermi level effects) strongly underestimates the diffusion, the calculated profile (not shown) being nearly the same as the initial one (no significant diffusion). This proves that nonequilibrium defects are, as usual, involved for these high concentration phosphorus diffusion conditions [15].

4.4. Suggested explanation

As demonstrated by the various experiments described above, the presence of the *doped* SOD film and of the RTP anneal are simultaneously required for a significant diffusion enhancement. However it is stressed that this enhancement is not linked to the particular spin-on deposition process, since an analogous enhancement has been reported for RTP diffusion using highly P doped SiO_2 film deposited by conventional atmospheric pressure chemical vapor deposition [18]. The diffusion enhancement is thus likely attributed to some transient phenomena associated with the phosphorus injection from the highly P doped glassy source. This assumption is further

strengthened by our recent observation on the influence of UV irradiation which causes a stronger diffusion enhancement because of a higher P injection [19].

It is now well accepted that phosphorus diffusion in Si occurs by a pair diffusion mechanism, in which the isolated substitutional P atoms are immobile, and only phosphorus/vacancy (PV) and phosphorus/self-interstitial (PI) pairs are mobile [15]. In equilibrium conditions, these pairs form by simple reactions between substitutional P and the isolated point defects (either V or I). In ATHENA, these pairs are assumed to be instantaneously in local equilibrium with the isolated species, and obviously this assumption is good enough for long time conventional furnace anneals. However, consider the case where the highly doped glassy source maintains a very high phosphorus chemical potential at the Si surface. Because of that, phosphorus is injected into the substrate and there is no reason why the P atoms should be instantaneously on substitutional sites. It is more than likely that a significant part of these atoms are injected as interstitials (or equivalently PI pairs). These PI's are in excess to the concentration needed to maintain the local equilibrium with the free self-interstitials and substitutional atoms, and then they will dissociate. However this dissociation is not instantaneous, and thus extra PI's exist during a finite time, during which the diffusion is enhanced. If the time constant for the dissociation is not negligible with respect to the total diffusion anneal time, this phenomenon will lead to a significant diffusion enhancement, as observed during RTP treatments.

5. Conclusion

We have reported various dedicated experiments in order to elucidate the physical origin of the diffusion enhancement observed for P diffusion from SOD sources during RTP. From these experimental results, in connection with theoretical simulations, we can rule out some hypothesis. It is suggested that the enhancement is due to a noninstantaneous dissociation of the PI pairs injected into the Si substrate by the presence of the highly doped SOD source.

Acknowledgements

This work was partially funded by ADEME-ECOTECH and by CEE in contract JOU3-CT95-0069.

References

[1] Wilson SR, Paulson WM, Gregory RB, Hamdi AH, MacDaniel FD. J Appl Phys 1984;55:4162.

[2] Sedgwick TO, Michel AE, Deline VR, Cohen SA, Lasky JB. J Appl Phys 1988;63:1452.

[3] Borisenko VE, Larsen AN. Appl Phys Lett 1983;43:582.

[4] Fogarassy E., Stuck R., Hodeau M., Williaux A., Toulemonde M., Siffert P. In: Platz, W. editor. Proceedings of the Photovoltaic Solar Energy Conference III, Cannes, France, 1980. , Dordrecht: Reidel Publ. Comp., 1981. p. 639.

[5] Carey PG, Sigmon TW, Press RL, Fahlen TS. IEEE Electron Dev Lett EDL 1986;7:440.

[6] Elliq M, Slaoui A, Fogarassy E, Pattyn H, Stuck R, Siffert P. Mater Res Soc Symp Proc 1991;109:159.

[7] Hartiti B, Slaoui A, Muller JC, Stuck R, Siffert P. J Appl Phys 1992;71:5474.

[8] Schindler R, Reis I, Wagner B, Eyer A, Lautenschlager H, Schetter C, Warta W, Hartiti B, Muller JC, Siffert P. Mater Res Soc Symp Proc 1991;109:162.

[9] Usami A, Ando M, Tsunckane M, Wada T. IEEE Trans Electron Dev TED 1992;39:105.

[10] Zagozdzon-Wosik W, Grabiec PB, Lux G. J Appl Phys 1994;75:337.

[11] Mar KM. J Electrochem Soc 1979;126:1252.

[12] Teh ST, Chuah DGS. Sol Energ Mater 1989;19:137.

[13] Usami A., Ando M., Tsunekane M., Yamamoto K., Wada T., Inoue Y. In: 18th IEEE Photovoltaïc Specialists Conf., 1985. p. 797.

[14] Itoh S, Homma Y, Sasaki E, Uchimura S, Morishima H. J Electrochem Soc 1990;137:1212.

[15] Mathiot D, Pfister JC. J Appl Phys 1984;55:3518.

[16] Reindl A. Solid-State Electron 1973;16:181.

[17] Mathiot D. Simulation Standard (SILVACO) 1996;7(2):1.

[18] Sivothaman S., Hartiti B., Nijs J., Barhdadi A., Rodot M., Muller J.C., Laureys W., Sarti D. In: 12th European Photovoltaic Solar Energy Conference, Amsterdam, 1994. p. 47.

[19] Noël S, Ventura L, Slaoui A, Muller JC, Groh B, Schindler R, Fröschle B, Theiler T. Appl Phys Lett 1998;72:2583.

PERGAMON

Materials Science in Semiconductor Processing 1 (1998) 237–241

MATERIALS
SCIENCE IN
SEMICONDUCTOR
PROCESSING

Ultra-shallow junction formation by spike annealing in a lamp-based or hot-walled rapid thermal annealing system: effect of ramp-up rate

Aditya Agarwal[a],*, Anthony T. Fiory[b], Hans-Joachim L. Gossmann[b],
Conor S. Rafferty[b], Peter Frisella[c]

[a]*Semiconductor Equipment Operations, Eaton Corporation, Beverly, MA 01915, USA*
[b]*Bell Laboratories, Lucent Technologies, Murray Hill, NJ 07974, USA*
[c]*Thermal Processing Systems, Eaton Corporation, Beverly, MA 01915, USA*

Abstract

Ultra-shallow p-type junction formation has been investigated using 1050°C spike anneals in lamp-based and hot-walled rapid thermal processing (RTP) systems. A spike anneal may be characterized by a fast ramp-up to temperature with only a fraction of a second soak-time at temperature. The effects of the ramp-up rate during a spike anneal on junction depth and sheet resistance were measured for rates of 40, 70 and 155°C/s in a lamp-based RTP, and for 50 and 85°C/s in a hot-walled RTP. B^+ implants of 0.5, 2 and 5 keV at doses of 2×10^{14} and 2×10^{15} cm^{-2} were annealed. A significant reduction in junction depth was observed at the highest ramp-up rate for the shallower 0.5-keV B implants, while only a marginal improvement was observed for 2- and 5-keV implants. It is concluded that high ramp-up rates can achieve the desired ultra-shallow junctions with low sheet resistance but only when used in combination with spike anneals and the lowest energy implants. © 1999 Elsevier Science Ltd. All rights reserved.

1. Introduction

Future device technology requires the ability to form extremely shallow junctions which simultaneously have a very low sheet resistance [1]. Though junction depth, x_j, can be limited by annealing for a sufficiently short time at a greatly reduced temperature, e.g. 750°C/10 s, this results in decreased dopant activation and insufficient damage removal leading to a large sheet resistance, R_s, and increased junction leakage, respectively. It is thus necessary to anneal at the highest temperature possible while limiting the total thermal budget for dopant diffusion. The thermal budget can be minimized by increasing the temperature ramp-up rate, and at the same time by decreasing the annealing time, to the limits allowed by the annealing equipment. Such

an anneal is referred to as a *spike* anneal in contrast to a *soak* standard rapid thermal anneal (RTA), where the ramp-up rate of ≈50°C is not maximized and the soak-time is a few seconds.

The dopant implantation process inevitably results in implantation-induced damage, i.e. Frenkel pairs consisting of vacancies and interstitials. During the initial stage of annealing most of the vacancies and interstitials recombine leaving behind a net excess of interstitials approximately equal to the implanted ion dose, the ' + 1' approximation [2]. These excess interstitials quickly coalesce into extended defects. The extended defects are metastable however and subsequently dissolve. As long as extended defects exist, an interstitial supersaturation is maintained, resulting in a boron diffusion enhancement. The enhancement ends soon after the defects have dissolved; the diffusion enhancement is thus transient. A consideration of the interstitial supersaturation and of the time to defect

---

dissolution [3] allows the increase in junction depth, Δx_j, due to dissolution of interstitial-type defects, to be expressed as [4]

$$\Delta x_j^2 \propto R_p \cdot \exp[-(-E_f + E_B - E_m)], \qquad (1)$$

where R_p is the projected ion range, $E_f + E_m = 4.9 \pm 0.1$ eV, is the activation energy for Si self diffusion and $E_B = 3.46$ eV [5], is the energy barrier for B diffusion. The activation energy of Δx_j is therefore ≈ -1.4 eV when boron diffusion is dominated by interstitial-type defect dissolution [4]. The linear dependence of TED on R_p, predicted by Eq. (1), was recently demonstrated experimentally [6, 7] and is the reason why lower energy implants result in reduced TED. The second prediction of Eq. (1), the negative activation energy of Δx_j, means that if TED is allowed to run its course at 750°C, the resulting junction will be deeper than if it was formed at 1000°C. This is one of the main reasons that RTA was adopted as a replacement for traditional furnace annealing for junction activation. The negative activation energy of TED is also the most compelling argument for ramping-up the temperature at a very high rate; the less time spent at the lower temperature, the less enhanced diffusion of boron will occur. Since the first report of ramp-up rates as high as 400°C/s [8] there has been an enormous interest in the industry in determining the technology node [1] at which high ramp rates will be essential for forming sufficiently shallow junctions with low R_s.

In this work we have investigated the effect of ramp-up rate on junction parameters such as R_s and x_j for various combinations of B energy and doses. An analysis of diffusion enhancement during spike annealing will be reported elsewhere [9].

2. Experimental

B was implanted into 150 mm n-type Si wafers to a high dose of 2×10^{15} cm^{-2} at 0.5, 2 or 5 keV; and to a low dose of 2×10^{14} cm^{-2} at 0.5 and 5 keV. The wafers were then annealed in an AG Associates HP 8108 lamp-based RTA at ramp-up rates of 40, 70 and 155°C/s and in an Eaton Reliance 850 hot-walled furnace[1] at 55 and 85°C/s[2]. The annealing recipes were especially designed to achieve spike annealing. Standard

---

[1] The hot-walled furnace consists of a SiC or quartz bell jar, which is continuously resistively-heated at the top, establishing a thermal gradient from top to bottom. A wafer is heated by raising it to the appropriate height, while high ramp-up rates are achieved by varying its height rapidly.

[2] The ramp-up rate was calculated using the time from 700 to 1050°C. Instantaneous rates during the 85°C/s ramp-up exceeded 100°C s.

3 s soak anneals were also done in the lamp-based RTA with the maximum ramp-up rate of 155°C/s. The ramp-down rate was constant for all anneals at ≈ 70°C/ s in the lamp-based RTA or ≈ 60°C/s in the hot-walled RTA. After annealing, all wafers were R_s-mapped. R_s uniformity ($1 - \Sigma$standard deviation) ranged from 0.7 to 9.5% for the lamp-based system and 0.5 to 4% for the hot-walled system. No effort was made in this experiment to improve uniformity on either RTA system since the goal was to achieve high ramp rates with the shortest anneal time. Pieces from selected wafers were analyzed by secondary ion mass spectroscopy (SIMS). Care was taken to analyze the portion of the wafer where the temperature was measured by the pyrometer and R_s values reported here correspond to the same portion.

3. Results and discussion

The R_s data for all spike-annealed wafers are shown in Fig. 1a. For all energy and dose combinations R_s increases with the ramp-up rate, regardless of the furnace type. It is interesting to note that the percent increase in R_s (Fig. 1b) is greater when the ramp-up rate is increased from 70 to 155°C/s than when it is increased from 40 to 70°C/s. Though the reduction in thermal budget by increasing the ramp-up rate from 70 to 155°C/s is smaller, the penalty imposed by the increasing sheet resistance is larger.

For both types of furnaces higher ramp-up rates lead to shallower junctions for the high B dose of 2×10^{15} cm^{-2} at 0.5 keV (Fig. 2a). The hot-wall annealed profiles are slightly deeper than the comparable rate profiles in the lamp-based system (Fig. 2a), consistent with the lower R_s values (Fig. 1a). A comparison of the lamp-based data for different energies reveals that the reduction in x_j achieved by increasing the ramp-up rate from 40 to 70 to 155°C/s is greater for the shallower 0.5-keV implant (Fig. 2a) than for the deeper 2-keV implant (Fig. 2b). The same trend is also observed for the lower B dose of 2×10^{14} cm^{-2} (Fig. 3); while the high ramp-up rate of 155°C leads to a shallower junction for the 0.5-keV implant, no difference can be discerned in the 5-keV profiles for different ramp-up rates. Note that the larger percentage increase in R_s with an increase in ramp-up rate from 70 to 155°C/s (Fig. 1b) is not accompanied by a larger decrease in x_j (Fig. 2a and b). Indeed x_j decreases less from 70 to 155°C/s than from 40 to 70°C/s, for both 0.5 and 2 keV B.

The data in Figs. 1–3 can be summarized by considering R_s as a function of x_j (Fig. 4). The data from this work are grouped by the B energy and dose combination. The trend pointed out in the previous section of diminishing advantage from a high ramp-up rate

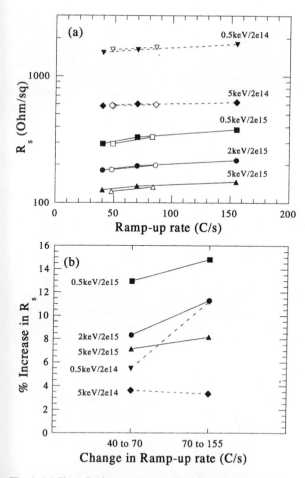

Fig. 1. (a) Sheet Resistance, R_s, as a function of ramp-up rate for various energy and dose combination B implants after 1050°C spike annealing. Anneals at 40, 70 and 155°C/s ramp-up rates were carried out in a lamp-based RTA and at 50 and 85°C/s in a furnace-based RTA. (b) Percentage increase in R_s for a change in ramp-up rate from 40 to 70°C/s and 70–155°C/s.

with increasing implant energy is also apparent in Fig. 4 where the vertical and lateral spread (R_s and x_j, respectively) between data points corresponding to different ramp-up rates decreases with increasing implant energy, for both furnace types. In other words, the impact of higher ramp-up rates decreases with increasing implant energy.

Data points for other annealing conditions are also included in Fig. 4 to help determine the relative advantages of spike anneals over standard anneals:

1. Soak time at high ramp-up rates: if the soak time for the high dose 0.5-keV B implant is increased from 0 to 3 s, x_j increases from 45 to 77 nm (Fig. 2a and Fig. 4). Moreover, despite having had the highest ramp-up rate of 155°C/s, after a 3 s anneal the

x_j and R_s combination for the 0.5-keV implant is not significantly different than for the higher energy 2-keV implant. It appears then that the advantage of high ramp-up rate for a low energy 0.5-keV implant can be fully realized only by a spike anneal, i.e. with soak time $\ll 1$ s. It can also be said that the advantage gained by reducing the implant energy can be compromised by performing a standard anneal instead of a spike anneal. The large increase in x_j upon annealing for 3 s also raises concerns about the difficulty in controlling time during the spike anneal process.

2. Lower temperature standard anneals: also shown in Fig. 4 is that a standard 1000°C, 10 s anneal of 2 keV B leads to x_j and R_s values which are very similar to those from a 1050°C spike anneal. These results imply that for 2 keV B a standard soak anneal at a reduced temperature can lead to the same junction properties as a more aggressive spike anneal.

Fig. 4 also includes data points from a process optimization experiment which included a preamorphization step and spike anneals with ramp-up rates as high

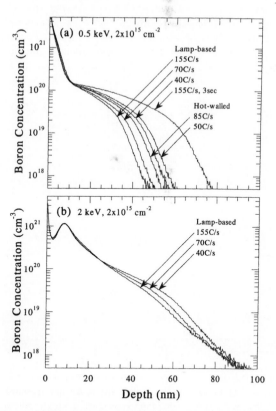

Fig. 2. SIMS B profiles as function of ramp-up rate for 1050°C spike anneals (a) for 0.5 keV, 2×10^{15} cm^{-2} B in a lamp-based and hot-walled RTA and (b) 2 keV, 2×10^{15} cm^{-2} B in a lamp-based RTA.

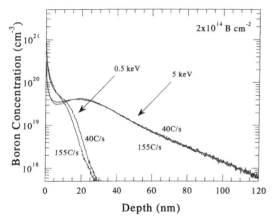

Fig. 3. SIMS B profiles for 2×10^{14} cm^{-2} B at 0.5 and 5 keV after 1050°C lamp-based spike RTA with ramp-up rates of 40 or 155°C/s.

as 400°C [10]. The data from Ref. [10] are quite similar to the data from this work, making it apparent that different implant and annealing options are available for achieving comparable shallow junction depths and sheet resistance. Our experiment does not however measure leakage current, which is expected to vary with residual damage following annealing. The ramp-up rate is only one variable which affects the junction. The complete list of variables must include implant species, energy and dose; preamorphization species, E,

and dose (if any); annealing temperature and time; annealing ambient and postprocessing temperature and time. Of course, the junction anneal process must be optimized in concert with the impact of temperature, time, ambient and ramp-rates on the n-type junction.

However, despite the large number of contributors to the junction properties, it is still possible to isolate the contribution of the ramp-up rate as has been done here. From the data presented here it is apparent that a high ramp-up rate can significantly reduce x_j and simultaneously achieve low R_s values for 0.5-keV B implants. The improvement decreases rapidly however with increasing implant energy.

The first of two parameters we use to describe the spike anneal is the peak temperature during the anneal. In our experiment the standard deviation in peak temperature increased from < 1°C at the lowest ramp-up rate of 40°C/s to > 6°C at the highest rate of 155°C/s. It is unknown at this time how much this run-to-run repeatability can be improved by optimizing the control system or even what the standard deviation corresponds to in process nonuniformity. It is, however, clear that the repeatability would only degrade with higher ramp-up rates.

The second metric which we have used is the time spent above 1000°C, plotted in Fig. 5 for all anneals in both furnaces. The time above 1000°C for spike anneals in the hot-wall furnace is ≈ 1 s longer than in the lamp-based system explaining both the slightly deeper junctions and the better uniformity. As with the peak temperature standard deviation in time also increases with ramp-up rate. For the highest rate of 155°C/s increasing soak time to 3 s reduces the

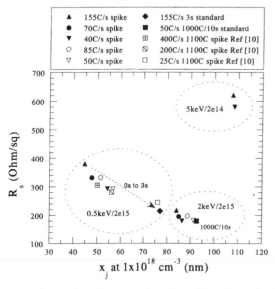

Fig. 4. Comparison R_s vs. x_j data for spike and standard anneals from this work and from Ref. [10], in which 10^{15} cm^{-2} 1 keV B was preceded by preamorphization with 10^{15} cm^{-2} 5 keV Ge. Ramp-up rates of 155, 70 and 40°C/s correspond to the lamp-based system, while 85 and 50°C correspond to the hot-walled system.

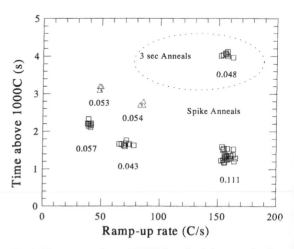

Fig. 5. Time spent above 1000°C for all of the anneals (data from the dummy wafers run prior to each implanted wafer). The standard deviation in time (s) is noted for each type of anneal.

standard deviation. As expected, control of both time and temperature degrade with increasing ramp-up rate. It remains to be seen how reliable and repeatable a spike anneal can be made.

4. Summary and conclusions

We have carried out an experiment to determine the effect of the ramp-up rate during spike annealing on the formation of ultra-shallow junctions in a lamp-based RTP and also for the first time in a hot-walled RTP system. Using various implantation energies and doses, it has been demonstrated that while there is a significant advantage from a higher ramp-up rate during spike annealing for low energy 0.5 keV implants, the advantage is comparatively marginal for higher energy 2-keV implants. It is also observed that the degradation in sheet resistance with increasing ramp-up rates is greater than the improvement in junction depth.

Acknowledgement

We thank John Jackson and Michael Walsh of Eaton Implant Systems Division for the low energy B implants.

References

[1] National technology roadmap for semiconductors. San Jose: Semiconductor Industry Association, 1997.

[2] Giles MD. J Electrochem Soc 1991;138:138.

[3] Rafferty CS, Gilmer GH, Jaraiz M, Eaglesham DJ, Gossmann H-J. Appl Phys Lett 1996;68:2395.

[4] Gossmann H.-J. In: Huff, H.R., Goselle, U., Tsuya, H., editors. Semiconductor silicon. ECS Proc. 98(1), 1998. p. 884.

[5] Fair RB, Tsai JCC. J Electrochem Soc 1977;124:1107.

[6] Agarwal A, Gossmann H-J, Eaglesham DJ, Pelaz L, Herner SB, Jacobson DC et al. IEDM Tech Digest 1997;0:367.

[7] Agarwal A, Gossmann H-J, Eaglesham DJ, Pelaz L, Jacobson DC, Haynes TE et al. Appl Phys Lett 1997;71:3141.

[8] Shishiguchi S, Mineji A, Hayashi T, Saito S. VLSI Tech Symp 1997;1059.

[9] Agarwal A., Fiory A.T., Gossmann H.-J., Rafferty C.S., Frisella P., Hebb J., Jackson J., Appl Phys Lett, submitted for publication.

[10] Saito S. Paper presented at MRS Spring 1998 Meeting, San Francisco.

PERGAMON

Materials Science in Semiconductor Processing 1 (1998) 243–247

MATERIALS
SCIENCE IN
SEMICONDUCTOR
PROCESSING

Formation of contacts to shallow junctions using titanium silicide with diffusion barriers

W. Zagozdzon-Wosik [a],*, I. Rusakova [b], S.R. Gooty [a], D. Marton [c], J. Li [a],
Z.H. Zhang [b], C.-H. Lin [d], R.J. Bleiler [e], D.X. Zhang [d]

[a]*Electrical and Computer Engineering Department, University of Houston, 4800 Calhoun, Houston, TX 77204, USA*
[b]*Texas Center for Superconductivity, University of Houston, 4800 Calhoun, Houston, TX 77204, USA*
[c]*Chemistry Department, University of Houston, 4800 Calhoun, Houston, TX 77204, USA*
[d]*Space Vacuum Epitaxial Center, University of Houston, 4800 Calhoun, Houston, TX 77204, USA*
[e]*Evans Texas, Austin, TX 78754, USA*

Abstract

We investigated the process integration of shallow junctions' formation and titanium silicidation. Junctions were doped with boron or arsenic during diffusion in rapid thermal processing (RTP). Dopant sources were in the form of pure layers of B or As deposited on the silicon substrate by e-beam evaporation or molecular beam epitaxy (MBE), respectively. Contact formation to the junctions was preceded by Ti deposition, either on a sacrificial layer of amorphous silicon or directly on the dopant layer. We studied the role of dopant–metal compound formation as a diffusion barrier to prevent junction degradation during the silicidation process. Such a barrier completely stops silicide formation in the case of B but is less efficient for the As doping. It also impedes boron diffusion into the silicon substrate much more than arsenic diffusion. A standard process sequence, where the junction is formed first and silicidation follows, and a modified process flow, where silicidation is done in situ together with the junction doping lead to differences in the fabricated structures. Various analytical techniques such as secondary ion mass spectroscopy (SIMS), Auger electron spectroscopy (AES), Rutherford backscattering spectroscopy (RBS), spreading resistance profiling (SRP), cross-section transmission electron microscopy (XTEM) and four-point probe were used for junction characterization. © 1999 Elsevier Science Ltd. All rights reserved.

1. Introduction

One of the fabrication challenges in the very large scale integration (VLSI) technology for integrated circuits (IC) is the shallow junctions' formation. Requirements for scaled down devices become more stringent in regard to dopant profiles with high surface concentrations and junction depths in the range of deep submicrometers. Many different doping techniques such as gas immersion laser doping (GILD) [1], gas [2] and solid state [3] diffusion processes and plasma immersion ion implantation [4] are being successfully studied to produce ultra shallow junctions at depths down to 100 Å^3. However, the development of doping techniques alone is not enough for device fabrication since ultra shallow junctions can be easily degraded by silicide contacts [5], either during the silicidation process, because of the Si consumption or in subsequent back-end thermal processes [6] such as the BPSG reflow. Dopant depletion at the junction surface, especially severe in the case of B, further deteriorates the contacts by increasing their resistance [7]. Process integration of doping and metallization plays therefore an important role in the shallow junction technology for scaled-down devices. It has been already addressed extensively by studying silicide as diffusion sources (SADS) [5, 8] for various types of silicides and process sequences, but its implementation in the IC manufacturing is still limited.

* Corresponding author.

Our goal in this paper is to address both aspects of downsizing vertical dimensions of the doped layers e.g. shallow junctions and contacts formation. Junction fabrication is realized by a diffusion process of boron or arsenic to obtain very small depths and ensure defect free silicon structure. Pure solid dopant layers (B and As) deposited directly on silicon wafers are used both for p^+ and n^+ junctions. Our earlier results show that annealing conditions in the rapid thermal processing readily control shallow dopant profiles in Si and sheet resistance (R_s) values [9] with little dependence on the dopant layer thickness. Contacts are made by the standard two step titanium silicidation process. To avoid junction degradation due to the Si substrate consumption, an amorphous silicon (α-Si) is introduced under the Ti layer to act as a sacrificial layer in silicidation, while diffusion barriers, formed on the substrate surface in a reaction between the respective (B or As) residual dopant and titanium, prevent silicide encroachment into the p^+ and n^+ junctions. Boron and arsenic can react with Ti to form metal–dopant compounds. TiB_2 is known as a good electrical conductor and diffusion barrier [10].

2. Experiment

Low doped n- and p-type (100) CZ Si wafers were used for B and As doping, respectively. Wafers were cleaned using IMEC recipe [11] with the HF last step, prior to the dopant source and/or Ti and/or α-Si deposition processes. Deposition of B layers was done by electron beam evaporation from outgassed pure melted 99.9% boron lumps. Arsenic layers were deposited by molecular beam epitaxy (MBE), instead of e-beam due to safety, and were always capped with a thin layer of Si to decrease dopant loss by evaporation during subsequent RTP. Deposition of a sacrificial α-Si layer and Ti film was done for B and As doping experiments by e-beam at a base pressure in the range of low 10^{-6} torr. Titanium was always protected against oxidation, due to residual oxygen in the ambient gas in RTP, by another α-Si layer which was only 70 Å thick and was also deposited in situ by e-beam evaporation from a target of low doped Si.

We used the following dopant sources on silicon: (a) a plain dopant layer, (b) dopant/Ti stack, (c) dopant/α-Si cap and (d) stack of B, α-Si cap and Ti layers, to test the role of dopants in Ti silicidation process via RTP. Annealing was done in the N_2 ambient using thermal conditions that either correspond to the diffusion processes from 850 to 1050°C for 10 s and/or to silicidation from 600 to 700°C for 90 s followed by 800–950°C for 20 s to form the C-54 phase of $TiSi_2$. Samples were analyzed using sheet resistance (R_s) measurements, Auger electron spectroscopy with depth

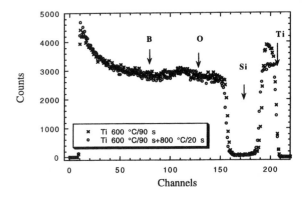

Fig. 1. RBS spectra obtained for the two step silicidation process. Thermal processes, which for Si ensure formation of the Ti silicide, do not result in silicidation due to the boron presence.

profiling, Rutherford back scattering, secondary ion mass spectroscopy and cross-section transmission electron microscopy. To increase the depth resolution in RBS a grazing angle was used by placing a detector at 103° to the beam.

3. Results and discussion

Results of boron diffusion into Si from the B source followed by Ti deposition and RTP annealing indicate that a residual dopant layer, formed at the silicon surface, indeed reacts with Ti and acts as a diffusion barrier against junction silicidation. This effect is illustrated in Fig. 1 by the RBS spectra after B diffusion at 900°C for 10 s, Ti layer deposition and subsequent annealing that corresponds to silicidation processes at 600°C for 90 s with or without a subsequent 800°C/20 s step. Silicide formation is hindered by the B presence and the titanium peak decreases with temperature indicating its increasing transformation [10, 12]. Similar results of impeded silicidation were obtained for higher temperatures used both for the dopant diffusion and silicidation processes. AES spectra (Fig. 2) show that the substrate silicidation does not occur, but also that boron does not diffuse into the Ti layer. Nonuniformity of Ti distribution is observed indicating a local reaction of Ti with B and the same slope of concentration decay for both elements demonstrates the compound formation. Boron not only prevents substrate silicidation, but also reduces the efficiency of junction formation by stopping the supply of boron to the substrate due to the source depletion through a compound layer. Similar, but much less limited B outdiffusion was documented in the SADS processes [13] using $TiSi_2$. The sheet resistance (R_s) after RTP at 900°C for 10 s confirms that the silicide indeed was not formed ($R_s = 28\,\Omega/sq$) into

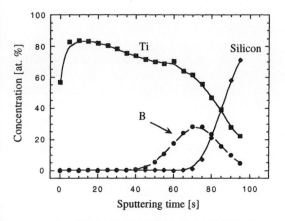

Fig. 2. AES results show nonuniformity of titanium distribution and lack of silicidation in the sample from Fig. 1. High concentration boron layer is present at the junction surface.

its C54 morphological phase. R_s values obtained after removal of the unreacted Ti also indicate very poor junction doping ($R_s = 2000\Omega$/sq) as compared with doping at the same thermal conditions from the plain B layer ($R_s = 170\Omega$/sq).

Similar effects were observed for the As doping process but the silicidation hindrance was not as complete as for boron. This is in contradiction to the smaller

silicide thickness [6] obtained for Si highly doped with As than with B. RBS results (Fig. 3) show that the as-deposited Ti layer reacts with the α-Si layer used as the capping layer for As, and with the second α-Si e-beam evaporated layer used as a Ti capping layer. The substrate silicon is shifted only negligibly towards lower energies. The silicide formation occurs, as indicated by the AES spectra (not shown here) but results in a titanium-rich silicide. Sheet resistance of such contacts is smaller compared to the As deposited films, but it is unacceptably high for the device application.

Unlike in the boron doping case, As is not completely confined to the titanium layer but diffuses to the silicon substrate. Profiles of arsenic after the silicidation processes at the standard thermal annealing conditions for samples with the Ti layer and its 70 Å α-Si cap only but without the sacrificial α-Si barrier show dopant penetration into the substrate.

In order to form a silicide contact, a layer of amorphous silicon has to be present between the boron or arsenic source and titanium. Such a α-Si cap can be deposited on the dopant layer prior to the RTP diffusion or after the diffusion process. Dopant diffusion into the α-Si cap is faster than into the crystalline substrate due to grain boundary diffusion thus results in layers heavily doped either with B or As. This results in higher dopant concentration in the silicide layer formed on the doped Si cap as compared to the undoped cap (Fig. 4). SIMS profiles show also smaller

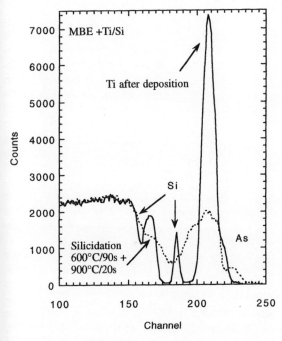

Fig. 3. RBS spectra obtained after silicidation process for As doping without a sacrificial Si cap. Arsenic diffuses to the surface, thin silicon layers react with Ti, but the substrate/silicide does not penetrate to the bulk.

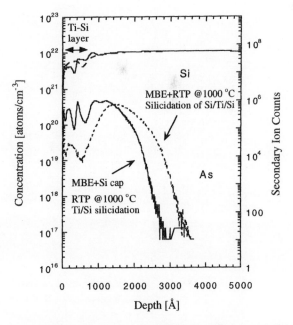

Fig. 4. SIMS profiles of As obtained for the diffusion and silicidation processes with the sacrificial Si cap. Sequence of the doping process and Si cap deposition plays an important role in dopant distribution in the cap and in the substrate.

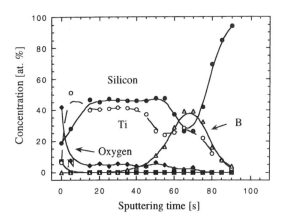

Fig. 5. AES spectra after silicidation for a sample with the Si sacrificial barrier deposited on the boron layer before RTP diffusion. High concentration of boron is present the silicon surface. Reaction with Ti leads to limitation in silicon supply from the substrate.

As penetration into the silicon substrate thus indicating the limited dopant supply.

When dopant diffusion from the deposited B or As layer is done prior to the silicide formation, it affects the subsequent silicidation process due to high concentration doping of the silicon cap. This is shown in Fig. 5 for B sources where AES profiles indicate for-

mation of a thick layer rich with Ti–B close to the interface of the dopant source and Si substrate and the titanium-rich silicide layer above. Interestingly, titanium profile is contained within the B-rich region, thus justifying the observed limitation of substrate silicidation. Distribution of oxygen proves that the thin protecting silicon cap used against oxygen contamination is effective in this RTP. Nonuniformity of the silicide composition observed in AES spectra are confirmed by XTEM analyses which show two distinct regions (Fig. 6), where the layer at the Si surface corresponds to the titanium–dopant compound-rich film and the upper one to titanium-rich silicide.

Another variant of the junction/contact formation process is one step annealing (600°C for 90 s followed by 900°C for 20 s), when doping of silicon substrate and silicidation of the amorphous cap occurs at the same time. It is possible to confine the silicidation to the cap layer, while the dopant–metal compound that forms at the Si surface acts as a diffusion barrier. Results for boron indicate that the formation of TiSi$_2$ takes place only within the cap layer and is limited by the boron layer. However, such a confinement of silicidation is less effective in the case of arsenic-doped junctions. In situ formation of the diffusion barriers both for n- and p-type-doped junctions during silicidation may open the possibility of process integration of

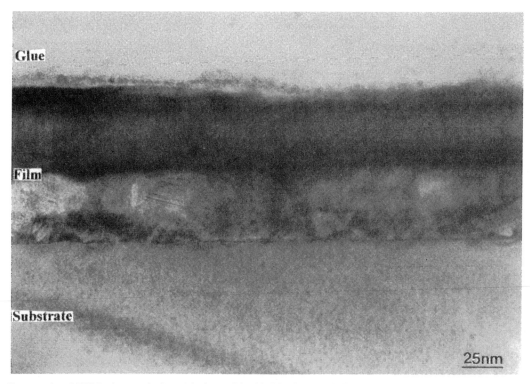

Fig. 6. Cross sectional TEM micrograph shows the layers identified in the sample from Fig. 5. Large grains correspond to high Ti concentration.

doping and metallization. Implementation of this concept in CMOS technology would rely on selective dopant deposition for NMOS and PMOS transistors and α-Si. This can be realized by substantial modification (ex. by GILD) of the present technology based on ion implantation.

Good quality of the silicided junctions, ensured by high dopant concentrations at the silicon surface, has been confirmed by measuring transmission line model test structures using 50 μm wide patterns. The contact resistance R_{cn} related to the unit width was about 0.104 cm and the corresponding specific contact resistance was 1×10^{-6} cm^2 even for the high temperature diffusion (1000°C/10 s) from the B/α-Si source followed by Ti silicidation.

The role of dopant in preventing junction silicidation may be perhaps useful also in other metallization schemes, when silicidation should be completely avoided, and refractory metallization be used to decrease RC constants of ICs [14]. Since metal dopant compound formation blocks silicide formation, shallow junctions may be protected and low resistances obtained.

4. Summary

In conclusion, we investigated the role of pure boron and arsenic layers, used as dopant sources in diffusion for p- and n-type junctions formed by RTP, in the Ti silicidation processes. The boron and arsenic layers are present at the silicon surface after the diffusion process and are used to form a diffusion barrier against substrate consumption in silicidation. It was shown that due to the boron–titanium compound formation, the silicidation could be completely blocked (for B), or diminished due to the As–Ti (for As) compounds. If a sacrificial α-Si cap is deposited below the Ti layer such a barrier allows confining the silicide to the α-Si cap region thus preventing junction degradation.

Acknowledgements

This work was sponsored by the Coordinating Board as an Advanced Research Program and is gratefully acknowledged. We would like to thank Dr. M. Strikowski for his help in e-beam processes.

References

[1] Weiner KH In: Fair RB, Lojek B, editors. 2nd International Rapid Thermal Processing Conference, RTP'94, 1994. p. 238–43.
[2] Kiyota Y, Onai T, Nakamura T, Inada T, Hirano Y. IEEE Trans Electron Devices 1992;39:2077.
[3] Iwai H, Momose HS, Saito M, Ono M, Katsumata Y. Microelectron Eng 1995;28:147.
[4] Qin S, McGrruer NE. Chan C, Warner K. IEEE Trans Electron Devices 1992;39:2354.
[5] Osburn CM In: Fair RB, editor. Rapid thermal processing. Academic Press, Inc., 1993.
[6] Georgiou GE, Abiko H, Baiocchi FA, Ha NT, NAkahara S. J Electrohem Soc 1994;141(5):1351.
[7] Van den hove L, Maex K, Hobbs L, Lippens P, De Keersmaecker R, Probst V, Schaber H. Appl Surf Sci 1989;38:430.
[8] Osburn CM, Wang QF, Kellam M. Canovai C, Smith PL, McGuire GE, Xiao ZG, Rozgonyi GA. Appl Surf Sci 1991;53:291.
[9] Zagozdzon-Wosik W, Korablev K, Rusakova I, Simons D, Shi JH, Chi P, Davis MF, Wolfe JC. In: Lojek B, Fair R, editors. Proceedings of the 3rd International Conference on Rapid Thermal Processing, RTP'95, 1995. p. 302.
[10] Cho CS, Ruggles GA, Osburn CM, Xing GC. J Electrochem Soc 1991;138:3053.
[11] Heyns M, Verhaverbeke S, Meuris M, Mertens P, Schmidt H, Kubota M., Philipossian A, Dillenbeck K, Graf D, Schnegg A, de Blank R, Higashi GS, Irene EA, Ohmi T, editors. Mat. Res. Soc. Symp. Proc. vol. 315, 1993. p. 35.
[12] Itoh H, Matsudaira T, Naka S, Hamamoto H. J Mat Sci 1989;24:420.
[13] Probst V, Schaber H, Mitwalsky A, Kabza H, Hoffmann B, Maex K, Van den hove L. J Appl Phys 1991;70:693.
[14] Hisamoto D, Nakamura K, Saito M, Kobayashi N, Kimura S, Nagai R, Nishida T, Takeda E. IEEE Trans Electron Device 1994;41:745.

PERGAMON

Materials Science in Semiconductor Processing 1 (1998) 249–255

MATERIALS
SCIENCE IN
SEMICONDUCTOR
PROCESSING

Effect of stress on silicide formation kinetics in thin film titanium–selicon system

Rajan V. Nagabushnam *, Rajiv K. Singh, Sujit Sharan [1]

Department of Materials Science and Engineering, University of Florida, Gainesville, FL 32603, USA

Abstract

As the CMOS device feature size scales below sub 0.5 μm, formation of low resistive C54 phase $TiSi_2$ becomes increasingly difficult. With the decreasing silicide thickness and shrinking line widths, unrealistic thermal budgets are required to transform the high resistive C49 phase to the low resistive C54 phase [Kittl J.A., Prinslow D.A., Apte P.P., Das M.F. Appl Phys Lett 1995;67:2308; Saenger K.L., Cabral Jr. C., Clevenger L.A., Roy R.A., Wind S. J Appl Phys 1995;78:7040; Clevenger L.A., Cabral Jr. C., Roy R.A., Lavoie C., Viswanathan R., Saenger K.L., Jordon-Sweet J., Morales G., Ludwig Jr. K.L., Stephenson G.B. Mater Res Soc Symp Proc 1996;402:257]. This phenomenon of sluggish C49 to C54 phase transformation, according to various researchers, has been attributed to the reduced nucleation density. i.e. relatively larger grained C49 phase formed as compared to the vertical and the lateral dimensions of the silicide. Motivated by this technologically important C49 to C54 phase transformation, we have studied the ways in which a *nonthermal* parameter such as the 'stress state', of the Ti–Si diffusion couple, could affect the silicide formation and the C49 to C54 phase transformation. The stress state of the Ti–Si diffusion couple was varied in two different ways: (i) stress state of the Ti film and (ii) stress state of the silicon surface.

The stress state of the Ti film was varied by changing the deposition parameters during sputtering. We have observed that thin titanium films, when deposited under compressive stress, result in a silicide film with a small grained C49 phase. Hence, an enhanced C49 to C54 phase transformation was observed for compressive Ti films as compared to tensile Ti films. The activation energies associated with the C49 to C54 transformation in these thin compressive Ti films are about 2 to 2.25 eV lower than those commonly reported in the literature [Kittl J.A., Prinslow D.A., Apte P.P., Das M.F. Appl Phys Lett 1995;67:2308; Saenger K.L., Cabral Jr. C., Clevenger L.A., Roy R.A., Wind S. J Appl Phys 1995;78:7040; Clevenger L.A., Cabral Jr. C., Roy R.A., Lavoie C., Viswanathan R., Saenger K.L., Jordon-Sweet J., Morales G., Ludwig Jr. K.L., Stephenson G.B. Mater Res Soc Symp Proc 1996;402:257; Matsubara Y., Horiuchi T., Okumura K. Appl Phys Lett 1993;62:2634; Mann R.W., Clevenger L.A. J Electrochem Soc 1994;141:1347; Ma Z., Allen L.H. Phys Rev B 1994;49:13501] for similar annealing conditions. The stress state of the silicon surface was varied by implanting low doses ($1e13/cm^2$) of three different species such as boron, arsenic and phosphorus. As in the case of compressive Ti films, the C49 to C54 transformation was found to be enhanced for the silicon surface with the lowest compressive stress. © 1999 Elsevier Science Ltd. All rights reserved.

---

* Corresponding author. Current address: Networking & Computing Systems Group, Motorola, Austin, TX-78731, USA.

[1] Process Development, Micron Semiconductor, Boise, ID-83705.

1. Introduction

Titanium silicide in its lowest resistive form has been used for both self-aligned silicidation and contact silicidation process for a long time. However this practice is being challenged by the ever decreasing feature size

Table 1
Measured stress levels of various titanium films used for our studies

Deposition type (A, B or C)	Substrate type	Titanium film thickness (Å)	Radius of curvature (m)		Stress (Mpa), T: +, C:−
			Pre	Post	
Type A	thermal oxide	500	83.75	96.46	−518.39
Type A	bare	1000	−168.07	−127.88	−308.0
Type A	bare	500	114.03	132.46	−402.17
Type A	bare	200	−168.71	−160.93	−236.0
Type B	thermal oxide	500	115.01	114.47	13.5
Type B	bare	500	94.82	91.03	144.49
Type B	bare	200	−604.74	−614.34	20.9

Table 2
Silicon substrate implantation details

Wafer	Species	Implantation dose $(1/cm^2)$	Implantation energy (keV)
Type IA	boron	1e13	50
Type IB	phosphorus	1e13	170
Type IC	arsenic	1e13	170

of the modern VLSI devices. It has been observed that with decreasing silicide film thicknesses required for the ultra shallow junction technology, the formation of low resistive C54 phase is becoming increasingly difficult [1–6]. In addition, various values have been reported either for the activation energy associated with the phase transformation or the C54 phase transformation temperature. However though these values differ by as high as 100% [1–6], they may not necessarily contradict each other as the process conditions used for each study were different from each other. Therefore a detailed analysis of certain fundamental parameters affecting the silicide growth and transformation is required. We therefore decided to study the effect of stress state of the thin titanium film-silicon couple on the C49 phase grain size and the subsequent transformation to C54 phase.

2. Experiment

Applied Materials Endura^R platform staging a RF (radio frequency) sputtering system was used to deposit titanium films from titanium target (99.9999% purity) for all the samples (6″ silicon wafers) used for our studies. The base pressure before the start of the deposition was maintained below 10^{-9} Torr and would increase to about 3×10^{-3} Torr during depo-

sition. The deposition temperature used varied from 200 to 300°C and the wafer temperature was maintained by heating the backside with heated ultrapure (99.9999%) argon.

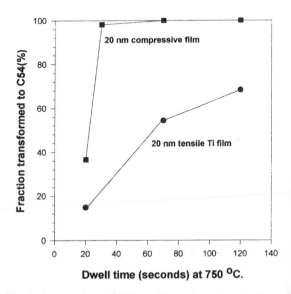

Fig. 1. A comparison of C49 to C5 transformation kinetics as observed in the Ti films of two different 'as deposited' stress levels.

In some instances a *collimator*, made of stainless steel, with an aspect ratio of 1.5:1 was used. The collimator has several openings of honey-comb shape, when placed between the target and the silicon substrate, would only allow the sputtered ions having normal or near normal incidence to reach the substrate. The actual purpose of this setup is to deposit films on the substrates having high density patterns and features having large aspect ratios. However this setup, as it reduces the flux of incident metal atoms on to the silicon substrate by physically blocking them, would require higher RF power to maintain substantial deposition rates. The stress state of the deposited thin titanium film and the silicon surface was measured using commercially available Tencor[R] Flexus system. Either (i) AG Heatpulse[R] 210 or (ii) Applied's susceptor based metal annealing system were later used to anneal the films for our studies.

Various characterization techniques such as X-ray diffraction (Siemens[R] D5600), transmission electron microscopy (JEOL[R] 200 CX), Auger electron spectroscopy and secondary ion mass spectroscopy (PHI[R] 6600) were used to obtain information relating to phase, microstructure and composition of the silicide films during various stages of rapid thermal annealing. As a first order, throughout our studies, the most commonly used four point probe method (FPPM) was used to observe the silicide formation by monitoring the change in the sheet resistance of the film.

2.1. Stress state of the titanium film

Lightly p-doped ($1e13/cm^3$) wafers with (100) orientation were used for our studies. The wafers were cleaned with 100:1 dilute hydrofluoric acid followed by deionized water (DI water) rinsing and loaded into the sputtering system. On one set of wafers, thin titanium films of thicknesses ranging from 135 to 1000 Å were deposited using high RF powers at a deposition rate of 0.35 Å/KW·s in the presence of the collimator (referred hereafter as *collimated film* or 'type A'). The other set of wafers, thin titanium films of two different thicknesses, 200 and 500 Å, were deposited using low RF powers leading to deposition rate of 3.33 Å/KW·s, in the absence of the collimator (referred hereafter as *noncollimated film or 'type B'*). The temperature used during both the deposition cases was 200°C. As shown in the Table 1, 'type A' films were observed to be more compressive in nature whereas 'type B' films exhibited a state of tensile stress. Though an extensive systematic kinetic study was performed in the case of compressive Ti films, of different thicknesses, by annealing at temperatures ranging from 700 to 800°C, our work in the case of tensile film was limited only to the 200-Å thick film for the mere sake of comparison. This decision was based on the fact that sputtered films used in several previously published detailed works [1, 4, 6, 7] was of conventional in nature similar to the one used in our work.

2.2. Stress state of the silicon surface

First, an 100-Å thick furnace based thermal oxide was grown on lightly doped p-type ($1e13/cm^3$) wafers with (100) orientation. To induce different levels of stress in the silicon surface, the silicon surface was implanted either arsenic, boron or phosphorus — the most commonly used dopant species in silicon microelectronics. The complete details of the implantation process are given in Table 2. One set of wafers belonging to 'type IA' through 'type IC' were also given an implantation activation anneal. The annealing cycle involved either 20 s at 1000°C for boron and phosphorus implanted samples or 1050°C in the case of samples implanted with arsenic. Then the native oxide was stripped using dilute 100:1 HF and loaded into the Ti deposition system. The wafers belonging to 'type IA' through 'type IC' were also read for changes in radius of curvature to monitor the stress state of the silicon surface after every step so far. 200-Å thick titanium (type A) was then deposited on all the wafers and rapid annealed thermally annealed in nitrogen ambient for different times at temperatures ranging from 700 to 800°C.

3. Results and discussion

3.1. Stress state of the titanium film

The transformation of C49 phase to low resistive C54 phase was observed to be enhanced during the entire stage of annealing cycle, when the stress state of titanium film in the 'as deposited' condition was 'compressive' instead of 'tensile' (Fig. 1). It is to be remembered in context that Hara et al. [7] too had observed such enhanced C49 to C54 transformation in the case of collimated Ti film as compared to conventional Ti film and that too purely limited to the early stages of transformation. Hara et al. attributed the differences in the transformation characteristics primarily due to the differences in the Ti film texture in the 'as deposited' condition. Though our conventional Ti film had a texture similar to that of the film experimented with by Hara et al., the collimated Ti film had a substantially different texture (a mixture of (100) and (002) texture) compared to the one of Hara et al. (a mixture of (101) and (002) texture). Fig. 2 compares the microstructure of the C49 phase obtained in the films of two different 'as deposited' stress states. We observed that, by including several

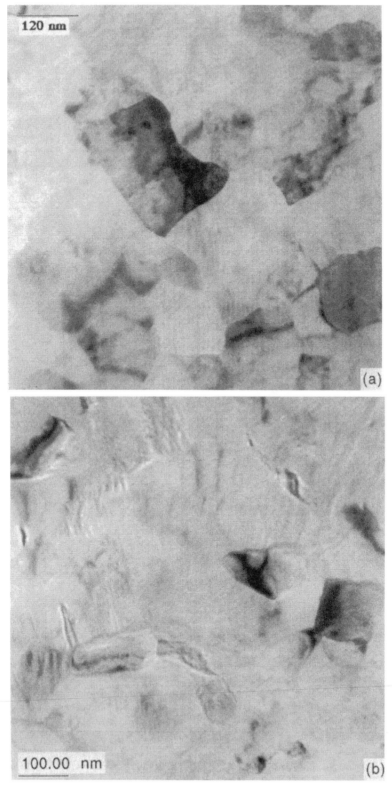

Fig. 2. TEM micrograph of C49 TiSl2 phase (i) Compressive Ti film (ii) Tensile Ti film.

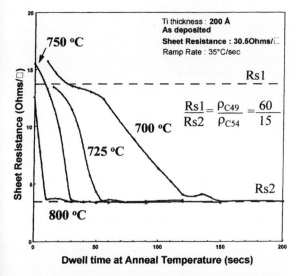

Fig. 3. Sheet resistance variation of 200-Å Ti–Si system with annealing for different annealing temperatures indicating the silicide formation and transformation to low resistive C54 phase.

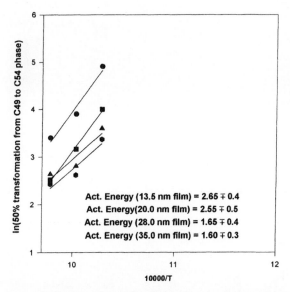

Fig. 5. Arrhenius plots of the time required to transform the silicide containing 100% C49 phase to a mixture containing 50% C49 and 50% C54 phase.

grains in our grain size analysis, the C49 phase grain size was about 26% smaller in the case of compressive film ($d = 126 \pm 20$ mn)as compared to tensile film ($d = 170 \pm 30$ nm).

The silicide formation and its subsequent rapid transformation to low resistive C54 phase during rapid thermal anneal cycle in the case of 200-Å compressive

Ti film is shown in Fig. 3. Similar trends were also observed in the case of films of other thicknesses. Characteristic sigmoidal curves were obtained in plotting the fraction of C49 phase transforming to C54 phase as a function of annealing time for different Ti film thicknesses (Fig. 4). These curves were indicative of a transformation involving nucleation and growth mechanism consisting of an incubation or induction period. As observed by others, the transformation in our work involving compressive Ti film too was

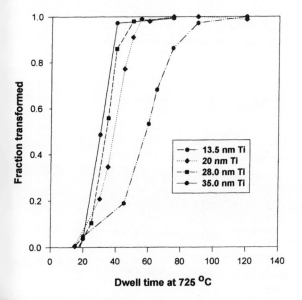

Fig. 4. Fraction of C49 phase transformed to C54 as a function of annealing time at 725°C for various Ti film thicknesses.

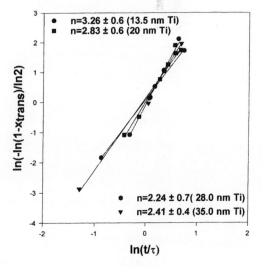

Fig. 6. Log–log plots for determining the mode of transformation for various Ti film thicknesses at 725°C.

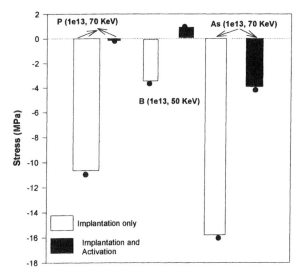

Fig. 7. Calculated surface stress levels in silicon samples receiving various surface treatments.

observed to be dependent on the original Ti film thickness i.e. thinner the Ti film, more difficult the transformation is. The effective activation energy, E_a, for the transformation in the case of compressive Ti film was extracted by plotting the natural log of the time at which 50% of the silicide film transformed to C54 phase (Fig. 5). As shown in the figure, activation energies ranging from 1.60 ± 0.3 to $2.65 + 0.4$ eV were obtained depending on the film thickness. These values are about 2–2.25 eV lower than what has been previously observed in the case of Ti films of different nature.

Adopting Johnson–mehl–Avrami approach for our nucleation mode analysis, Avrami coefficients ranging

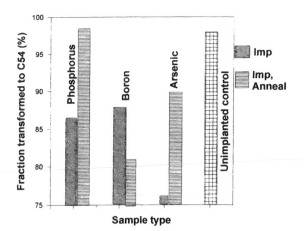

Fig. 8. Comparison of C49 to C54 transformation in various silicides formed on silicon substrates with different surface characteristics.

from 1.6 ± 0.4 (350 Å Ti) to 4.6 ± 0.7 (135 Å Ti) were obtained by plotting $\ln\{\ln[(1 - X_{\text{trans}})/\ln 2]\}$ versus $\ln(t/\tau)$ (Fig. 6). Though the values obtained for the films of thickness ≥ 200 Å are in good agreement with the already published values [1, 6], an high value of 3.26 ± 0.6 for the 135-Å thick film suggests that more nucleation sites are being generated as the transformation progresses. The only possible way for such phenomena to occur is when the top nitride–silicide interface starts acting as a surface of C54 nucleation sites, when the kinetic barriers for C54 phase are favorably lowered.

3.2. Stress state of the silicon surface

As Fig. 7 shows, the silicon surface exhibits a state of high compressive state right after the implantation step. This calculation was based on the assumption that the stress affected region is of the same order as the 'FWHM (full width at half maximum)' of the dopant profile. Of all the samples receiving only the implantation, the sample surface implanted with low dose born implant ($1e13/\text{cm}^3$, 50 keV) is at the least compressive state. The corresponding stress levels are higher for the samples with phosphorus and arsenic. The calculated stress level trend was in direct correlation with the C49 to C54 transformation observed in these samples (Fig. 8). In the 'as implanted' condition, the C49 to C54 transformation was highest in the boron implanted samples as compared to other samples. However, in the case of annealed samples, the sample surface implanted with phosphorus displayed the lowest compressive stress as compared to the sample implanted with arsenic or boron. The C49 to C54 transformation was also found to be most enhanced in these phosphorous implanted samples. We also observed that, in the case of phosphorus and arsenic implanted samples, the samples receiving the activation anneal after the implantation (i.e. silicon surface at a less compressive state) always exhibited smaller C49 grain compared to the samples which had only received the implantation. For instance, the C49 grain size in the case of arsenic implanted sample measured about 145 ± 35 nm and 200 ± 60 nm respectively for the two different silicon surfaces mentioned above.

4. Summary

In our work, the C49 to C54 transformation was found to be enhanced when (i) Ti film was deposited compressive or (ii) the state of the silicon surface was made tensile or less compressive. Relatively smaller C49 phase grain size was observed in the above cases leading to an enhanced C49 to C54 transformation kinetics. We propose that, based on our work, a slower or reduced movement of silicon from either a tensile silicon surface and/or through a compressive Ti film could lead to a relatively thinner amorphous Ti–Si mixture at the Ti–Si interface which could be potentially responsible for the small grained C49 phase.

References

[1] Kittl JA, Prinslow DA, Apte PP. Das MF. Appl. Phys. Lett. 1995;67:2308.

[2] Saenger KL, Cabral C, Jr, Clevenger LA, Roy RA, Wind S. J. Appl. Phys. 1995;78:7040.

[3] Clevenger LA, Cabral C, Jr, Roy RA, Lavoie C, Viswanathan R, Saenger KL, Jordon-Sweet J, Morales G, Ludwig KL, Jr, Stephenson GB. Mater. Res. Soc. Symp. Proc. 1996;402:257.

[4] Matsubara Y, Horiuchi T, Okumura K. Appl. Phys. Lett. 1993;62:2634.

[5] Mann RW, Clevenger LA. J. Electrochem. Soc. 1994;141:1347.

[6] Ma Z, Allen LH. Phys. Rev. B 1994;49:13501.

[7] Hara T, Nomura T, Mosley RC, Suzuki H, Sone K. J. Vac. Sci. Technol. A 1994;12:50.

PERGAMON

Materials Science in Semiconductor Processing 1 (1998) 257–261

MATERIALS
SCIENCE IN
SEMICONDUCTOR
PROCESSING

Strain relaxation and dopant distribution in the rapid thermal annealing of Co with Si/Si$_{1-x}$Ge$_x$/Si heterostructure

Y. Miron[a], M. Efrati Fastow[a], C. Cytermann[b], R. Brener[b], M. Eizenberg[a, b, *], M. Glück[c], H. Kibbel[c], U. König[c]

[a]*Department of Materials Engineering, Technion — Israel Institute of Technology, 32000 Haifa, Israel*
[b]*Solid State Institute, Technion — Israel Institute of Technology, 32000 Haifa, Israel*
[c]*Daimler–Benz AG, Ulm Research Center, D-89081 Ulm, Germany*

Abstract

The reaction of cobalt with the Si-sacrificial cap in the strained Si/Si$_{1-x}$Ge$_x$/Si MBE grown heterostructure was studied. The Si-cap is added to prevent the relaxation of the SiGe and to guarantee uniform and reliable silicidation reaction. The Si$_{1-x}$Ge$_x$ epilayer, with Ge content between 18 and 28 at%, was highly B doped, while the Si-cap was undoped or B doped either during growth or by ion implantation. Cobalt evaporation was followed by rapid thermal annealing at 450–700°C for 30 sec in N$_2$ or Ar + 10%H$_2$. When the silicide penetrated the Si-cap/Si$_{1-x}$Ge$_x$ interface, noticeable out-diffusion of Ge and B to the surface was observed. In spite of the presence of the Si-cap significant strain relaxation was observed in three cases: (1) in the implanted samples, although the implantation was confined to the Si-cap, (2) when the Co layer was too thick, such that the silicide penetrated the SiGe layer and (3) when the Ge content in the SiGe layer was relatively high (27.5%). © 1999 Elsevier Science Ltd. All rights reserved.

1. Introduction

Si$_{1-x}$Ge$_x$ heterostructures are attractive for electronic and optoelectronic devices due to their high mobility and band-gap engineering possibilities, while maintaining compatibility with the widely spread Si process technology.

CoSi$_2$ is one of the most promising silicides for the Si (and SiGe) based technology [1, 2]. The resistivity of CoSi$_2$ is very low and its formation temperature is relatively low as well. The direct reaction of a metal, and particularly Co, with the Si$_{1-x}$Ge$_x$ leads to significant strain relaxation at relatively low temperatures and to a delay of the reaction, requiring higher annealing temperatures for contact formation [2–5]. Therefore, the use of a sacrificial Si layer ('Si-cap') above the SiGe epilayer has been suggested to avoid

the direct reaction with SiGe [3, 6–8]. The thickness of the layers is restricted by the fact that each Å (in thickness) of Co reacts with 3.64 Å of Si to form 3.52 Å of CoSi$_2$.

Rapid thermal annealing (RTA) in inert gas ambient meets the requirements of low thermal budget and processing. No dependence of the inert gas environment was observed for CoSi$_2$ [6]. The phase sequence in the reaction of Co with Si is Co → Co$_2$Si → CoSi → CoSi$_2$ [9, 10]. The resistivity of these phases is ≅70, 100–150 and 14–17 μΩ·cm for Co$_2$Si, CoSi and CoSi$_2$, respectively [9].

When the SiGe is used as a base in a heterojunction bipolar transistor (HBT), it is usually highly doped. To form an ohmic contact, the sacrificial Si layer must be highly doped as well. Fig. 1 displays a schematic description of the CoSi$_2$ contact to the SiGe base, using a Si-cap layer. The doping of the Si-cap can be done during growth or by ion implantation. During thermal processing, the dopants may redistribute and

---

* Corresponding author.

1369-8001/99/$ - see front matter © 1999 Elsevier Science Ltd. All rights reserved.
PII: S1369-8001(98)00032-8

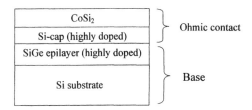

Fig. 1. A schematic description of the CoSi₂ ohmic contact to the SiGe base.

affect the Si–Co interaction. No compound formation between dopants and Co is reported* in the literature. Zaring et al. [11] found that B atoms in the consumed Si layer redistribute within the formed silicide layer and accumulate at the surface, while B in the underlying unconsumed Si is unaltered, and no accumulation due to snowplowing occurs at the silicide/Si interface.

In this work the reaction between Co and the Si-sacrificial layer is investigated. The Si sacrificial layer is undoped or B doped either during growth or by ion implantation. Special attention is given to the B redistribution and to strain relaxation of the SiGe epilayer due to applied heat treatments.

2. Experimental

The fully strained Si/Si$_{1-x}$Ge$_x$ epilayers were grown by molecular beam epitaxy (MBE) on (001) Si-substrates. In all samples, the 40 nm thick Si$_{1-x}$Ge$_x$ layer was in situ doped with B to a level of 4×10^{19} at/cm^3. The Si-cap was either undoped or B doped during growth or doped by BF$_2$ ion implantation as described in Table 1. The BF$_2$ implantation in samples B and D was confined within the Si-cap (Rp $\sim$ 50 nm, ΔRp $\sim$ 20 nm). Rapid thermal annealing at 900°C for 30 s in N$_2$ ambient followed the ion implantation. Prior to cobalt deposition the samples were cleaned in H$_2$SO$_4$:H$_2$O$_2$ (2:1) to remove the organic residuals and then dipped in diluted HF solution to remove the native oxide. Cobalt layers were deposited by e-beam evaporation in a vacuum of approximately 1×10^{-7} Torr. Silicidation was accomplished by rapid thermal annealing in N$_2$ or Ar + 10%H$_2$ flow for 30 s. Samples A and B were

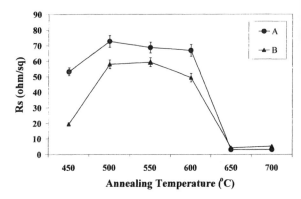

Fig. 2. Sheet resistance as a function of the annealing temperature for samples A (20 nm Co/100 nm Si-cap, in situ doped with (B) and B (20 nm Co/100 nm Si-cap, BF$_2$ implanted).

annealed at the temperature range of 450–700°C, while samples C and D at 600–700°C.

Sheet resistance was measured with a four-point-probe (4pp). Phase identification was done with X-ray powder diffractometry (XRD) in a Philips PW-1820 diffractometer (CuK$_\alpha$). High resolution four crystals X-ray diffractometer (HRXRD) was used to monitor the strain level of the SiGe epilayer (CuK$_\alpha$). Elemental analysis was achieved by AES (Physical Electronics 590A) and SIMS (Cameca IMS4f). Cross-sectional TEM (JEOL 2000 FX) was used to study the morphology of the layers.

3. Results and discussion

XRD measurements and AES depth profiles (not shown) of the samples of series A and B revealed the expected sequence of phase formation. After 450°C no reaction occurred. Between 500 and 600°C mainly CoSi was observed, but at 600°C CoSi₂ started to appear. The reaction was completed at 650°C with the formation of CoSi₂.

The sheet resistance of series A and B after 30 s RTA at 450–700°C is shown in Fig. 2. The shape of these curves can be directly correlated with the

Table 1
A detailed description of the samples layers and doping. The Ge content was calculated from HRXRD measurements

Name		Co (nm)	Si-cap (nm)		Si$_{1-x}$Ge (nm)	Ge (%)
A		20	100	B doped, 4×10^{19} at/cm^3	40	18.4
B		20	100	BF$_2$ implanted, 60 keV, 4×10^{15} at/cm^2	40	19.1
C	C1	18	60	undoped	40	27.5
	C2	10				
D		18	60	BF$_2$ implanted, 60 keV, 4×10^{15} at/cm^2	40	$\sim$27.5

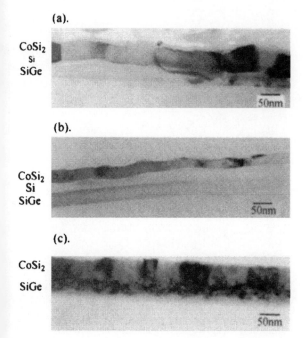

(a).

CoSi₂
Si
SiGe

50nm

(b).

CoSi₂
Si
SiGe

50nm

(c).

CoSi₂

SiGe

50nm

Fig. 3. Cross sectional TEM image of sample (a). C1 (18 mm Co/60 nm Si-cap) (b). C2 (10 nm Co/60 nm Si-cap) and (c). D (18 nm Co/60 nm Si-cap) after annealing at 700°C.

sequence of phase formation as indicated by XRD and AES results. At 450°C, the sheet resistance is related to the unreacted Co layer. The high resistance at 500–600°C can be explained by the high resistivity of the CoSi phase. Above 600°C a decrease in the sheet resistance indicates the transformation from the high resistivity CoSi to the low resistivity CoSi₂ phase. Finally, at 650–700°C only the low resistivity phase, CoSi₂, exists. The resistivity of CoSi₂ in sample A annealed at 650 and 700°C is 22.4 ± 1 and 21.9 ± 1 ($\mu\Omega \cdot$ cm), respectively, and that of sample B is 30.5 ± 1.4 and 37.6 ± 1.6 9$\mu\Omega \cdot$ cm), respectively. The resistivity found for set A agrees with values reported in literature, while that for B is somewhat higher. The higher resistivity may be attributed to a smaller grain size of the CoSi₂ as observed for samples C1 and D and will be discussed below.

Fig. 3 shows the cross sectional TEM images of samples C1, C2 and D after annealing at 700°C. In sample C1 (Fig. 3a), two regions are clearly observed: one in which a very thin layer of Si is still left between the CoSi₂ and the SiGe, and their interface is very smooth. In the other region there is no evidence of Si underneath the CoSi₂, and penetration of the silicide into the SiGe is clearly seen. In this region the interface is rough and defects are seen in the SiGe. The observation of two regions is an indication of non uniformity in the Co thickness or in the reaction rate (due, for example, to existence of native oxide at the

Co/Si interface). Fig. 3(b) shows the cross section of sample C2, with the thinner Co layer. A much thicker layer of Si is clearly seen between the silicide and the SiGe, and no penetration of the silicide into the SiGe is detected. In sample D (Fig. 3c) along the entire interface penetration into the SiGe, which contains many defects, can be observed. The interface with the SiGe is obviously very rough. The average grain size of the CoSi₂ in samples C (1 and 2) and D is approximately 100–150 and 30–50 nm respectively. Lur and Chen [12] also observed a similar difference in the grain size of CoSi₂ formed on implanted and nonimplanted samples.

SIMS depth profiles of sample A in the as-deposited state and after annealing at 700°C are shown in Fig. 4. It is obvious that the distribution of B and Ge in the unreacted layers is not altered due to the silicidation reaction. However, the B from the reacted Si-cap diffused through the silicide and accumulated on the surface. SIMS depth profiles taken from samples C1, in which the silicide penetrated the SiGe, are shown in Fig. 5. Segregation of Ge and B to the surface is observed at all annealing temperatures, but especially at 650 and 700°C. Since the Co–Si interaction is favored on the Co–Ge interaction [13], the consumption of the SiGe releases Ge that segregates to the surface. It also creates a route for grain boundary diffusion of B and Ge from the epilayer to the surface. A possible driving force for the out-diffusion of B may be the formation of B_2O_3 on the surface [11].

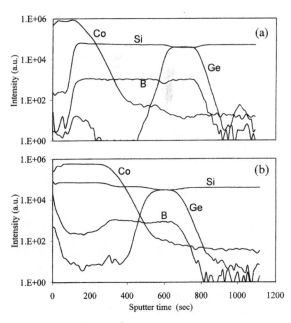

Fig. 4. SIMS profile of sample A (20 nm Co/100 nm Si-cap, in situ doped with B) (a) as-deposited and (b) after annealing at 700°C.

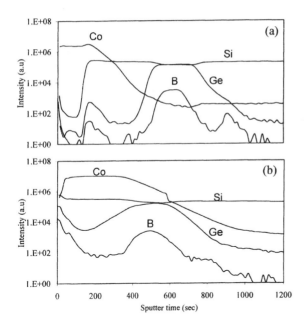

Fig. 5. SIMS profile of sample C1 (18 nm Co/60 nm Si-cap, undoped) in (a) as deposited and (b) after annealing at 700°C.

The strain level of the SiGe epilayers was derived from high-resolution XRD measurements of (004) and (115) rocking curves. The Ge content in the SiGe epilayer was derived using the experimental values of $Si_{1-x}Ge_x$ lattice parameter found by Dismukes et al. [14], and considering the compensation effect of the high concentration of B [7, 15]. It was found to be 18.4, 19.1 and 27.5% in samples A, B and C, respectively. Similar values were measured by AES.

Fig. 6 shows the strain level of the SiGe as a function of the annealing temperature. For sample C1 the strain level decreased, namely relaxation progressed with the annealing temperature. The penetration of the silicide into the SiGe, as observed by TEM, allows the nucleation and propagation of misfit dislocations. It should be noted that the strain measured by the HRXRD is an average over a very large area, and we can not resolve between the previously described two different zones that were observed in the TEM. For sample C2 (10 nm Co/60 nm Si-cap), the relaxation at 700°C was considerably lower than for sample C1 (18 nm Co/60 nm Si-cap). The relaxation in both samples occurred due to the relatively high concentration of Ge, however, in the thick Co samples, the penetration into the SiGe further contributed to the relaxation.

Sample A maintained its strain at all temperatures, while in sample B strain relaxation proceeded as the temperature increased. In sample B the silicide did not penetrate the SiGe, but nevertheless, strain relaxation was significant. Furthermore, the Ge content was relatively moderate (19.1%) and close to that in sample A.

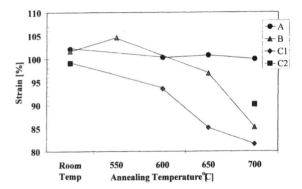

Fig. 6. Strain level as a function of the annealing temperature for samples A (20 nm Co/100 nm Si-cap, in situ doped with B), B (20 nm Co/100 nm Si-cap, BF$_2$ implanted), C1 (18 Co/ 60 nm Si-cap, undoped) and C2 (10 nm Co/60 nm Si-cap, undoped).

This extensive relaxation is attributed to the implantation damage. One might argue that it was caused by the activation annealing process that followed the ion implantation, however, the as-deposited sample (which has already been annealed after the implantation) was found to be fully strained. We believe that although the implantation was confined to the Si-cap layer, the damage in the layer was not fully annealed. When silicidation took place, residual damage, such as vacancies, probably diffused toward the Si/SiGe interface and enabled the nucleation of half loop dislocation, thus inducing strain relaxation in the SiGe epilayer. Further work must be done to fully understand this phenomenon.

4. Conclusions

When the silicidation reaction does not penetrate the SiGe, B atoms from the reacted zone diffuse through the silicide and accumulate on the surface, while the distribution of B and Ge in the underlying layers is unaltered by the reaction. When there is penetration into the SiGe, the B and Ge redistribute and both tend to accumulate on the surface, while decreasing their concentration in the underlying layers.

During the process of contact formation, the Si-cap can prevent strain relaxation of the underlying SiGe epilayer under the following restrictions: (1) the thickness of the deposited metal must be highly controlled, so that after silicidation the silicide will not penetrate into the SiGe, (2) the doping of the Si-cap should be done during the growth of the epilayer and not by BF$_2$ implantation, even if the implantation is confined within the Si-cap and (3) the Ge content in the SiGe should be kept moderate.

Acknowledgements

This research was supported by Daimler–Benz Aktiengesellschaft. Professor F. Meyer, Orsay University, France, is acknowledged for her help in rapid thermal annealing.

References

[1] Zaring C, Pisch A, Cardenas J, Gas P, Svensson BG. J. Appl. Phys. 1996;80:2742.

[2] Chen J, Colinge J-P, Flandre D, Gillon R, Raskin JP, Vanhoenacker D. J. Electrochem. Soc. 1997;177:2437.

[3] Nur O, Willander M, Radamson HH, Sardela MR, Jr, Hansson GV, Petersson CS, Maex K. Appl. Phys. Lett. 1994;64:440.

[4] Nur O, Willander M, Hultman L, Radamson HH, Hansson GV, Sardela MR, Jr, Greene JE. J. Appl. Phys. 1995;78:7063.

[5] Cytermann C, Holzman E, Brener R, Fastow M, Eizenberg M, Glück M, Kibbel H, König U. J. Appl. Phys. 1998;83:2019.

[6] Glück M, Schüppen A, Rösler M, Heinrich W, Hersener J, König U, Yam O, Cytermann C, Eizenberg M. Thin Solid Films 1995;270:549.

[7] Rodríquez A, Rodríquez T, Kling A, Soares JC, da Silva MF, Ballesteros C. J. Appl. Phys. 1997;82:2887.

[8] Donaton RA, Kolodinski S, Caymax M, Roussel P, Bender H, Brijs B, Maex K. Appl. Surf. Sci 1995;91:77.

[9] Van den Hove L, Wolters R, Maex K, De Keersmacher R, Declerck G. J. Vac. Sci. Technol. B 1986;4:1358.

[10] Maex K. Mater. Sci. Eng. 1993;R11:53.

[11] Zaring C, Svensson BG, Östling M. Mat. Res. Soc. Symp. Proc. 1992;260:157.

[12] Lur W, Chen LJ. J. Appl. Phys. 1988;64:3505.

[13] Wang Z, Aldrich DB, Chen YL, Sayers DE, Nemanich RJ. Thin Solid Films 1995;270:555.

[14] Dismukes JP, Ekstrom L, Paff RJ. J. Phys. Chem. 1994;68:3021.

[15] Sardela Jr MR, Radamson HH, Ekberg JO, Sundgren J-E, Hansson GV. Semicond. Sci. Technol. 1994;9:1272.

PERGAMON

Materials Science in Semiconductor Processing 1 (1998) 263–266

MATERIALS
SCIENCE IN
SEMICONDUCTOR
PROCESSING

Rapid thermal annealing of Zr/SiGeC contacts

M. Barthula[a],[*], V. Aubry-Fortuna[a], F. Meyer[a], O. Chaix-Pluchery[b],
A. Eyal[c], M. Eizenberg[c], P. Warren[d]

[a]*Institut d'Electronique Fondamentale, CNRS URA 22, Bât. 220, Université Paris Sud, 91405 Orsay Cedex, France*
[b]*LMGP, CNRS UMR 5628, ENSPG, BP75, 38402 St. Martin d'Hères, France*
[c]*Solid-State Institute, Technion, Haifa 32000, Israel*
[d]*Institute for Micro- and Optoelectronics, Swiss Federal Institute of Technology, 1015 Lausanne, Switzerland*

Abstract

In this work, the effect of rapid thermal annealing (400–800°C, 5 min) on the electrical properties of Zr/$Si_{1-x-y}Ge_xC_y$ contacts was investigated. Previously, we have shown that the reaction of Zr with SiGeC leads to the final compound C49-$Zr(Si_{1-z}Ge_z)_2$, with $z = x$ for all compositions examined and that no Ge-segregation is detected. After the reaction, only a small strain relaxation is observed in the unreacted SiGe epilayer, while the strain is totally preserved in the SiGeC one. Schottky barrier heights have been studied as a function of the annealing temperature. RTA leads to a decrease of the barrier for both n- and p-type. The decrease of the barrier height with reverse voltage is always well described by the thermionic emission. Nevertheless, the slight increase of the standard deviation, deduced from barrier height histograms, may show that few interface defects are created during annealing. © 1999 Elsevier Science Ltd. All rights reserved.

1. Introduction

Lately, there has been a growing interest in SiGe alloys for applications in electronic devices such as heterojunction bipolar transistors (HBT) and modulation-doped field effect transistor (ModFET). However, the 4.2% lattice mismatch between Si and Ge imposes severe restrictions in terms of critical thickness and thermal stability. Adding C to $Si_{1-x}Ge_x$ provides additional parameters in strain control [1,2] and bandgap engineering [3]. The fabrication of reliable ohmic and rectifying contacts is a critical step for new IV–IV heterostructure devices and requires a good control of electrical properties and thermal stability. Indeed, these structures are sensitive to any high-temperature treatment, which can lead to strain relaxation and (or) dopant diffusion. This results in an unacceptable degradation of the device performances. Some experiments have been carried out to study the reaction of

SiGe with Ti [4–6], Pd [7], W [8], Co [6,9,10] and other metals. The thermal treatments are usually performed with conventional annealing either under ultra high vacuum (UHV) [4,5,7] or under an atmosphere of inert gas [9]. Strain relaxation [5] of the epilayer and Ge-segregation [4,5,7,9] mostly occur during such thermal treatments. One way to prevent these detrimental effects is to reduce the thermal budget by using rapid thermal annealings (RTA) [6,8,10]. RTA is a convenient technology for production which also allows a good control of the reaction environment. Nevertheless, in the case of the Ti/SiGe system, Aldrich et al. [4] have shown that RTA only leads to a reduction of Ge-segregation but does not prevent it totally, and a dramatic strain relaxation has been reported after Co/SiGe reaction performed in a RTA system [6,10]. Therefore, it is also important to choose a metal which in principle avoids Ge-segregation. Recently, the Nemanich group [11] has proposed a new attractive system with Zr as a metal. Their experimental results showed that the system Zr/SiGe is more stable to Ge-segregation than the Ti/SiGe one,

* Corresponding author.

even after an annealing at 700°C for 100 min. In previous articles, we have investigated the thermal stability of the $Zr/Si_{1-x}Ge_x$ system using RTA at temperatures ranging from 400 to 800°C [12, 13]. The XRD analyses have indicated that $C49\text{-}Zr(Si_{1-y}Ge_y)_2$ with $y = x$ was the final phase of the $Zr/Si_{1-x}Ge_x$ reaction, regardless of the initial $Si_{1-x}Ge_x$ alloy compositions ($0.1 \leq x \leq 0.33$). At 600°C, the reaction forming the orthorhombic $Zr(Si_{1-y}Ge_y)$ has begun and the formation of $C49\text{-}Zr(Si_{1-y}Ge_y)_2$ is initiated. At 800°C, the phase $C49\text{-}Zr(Si_{1-y}Ge_y)_2$ is not always completely achieved, some $Zr(Si_{1-y}Ge_y)$ peaks can be still detected. According to XRD, RBS and EDS measurements, no Ge-segregation has occurred during the reaction. The TEM images confirm that no Ge-excess is detected at the interface, at the surface, nor at the grain boundaries. No modification of the reaction was observed due to C-incorporation. In addition, FCD analyses were performed to follow the state of strain of the unreacted epilayer [13]. A small strain relaxation has only occurred in the unreacted SiGe epilayer during the reaction, while the strain is totally preserved in the SiGeC one. All these results lead to the conclusion that Zr offers the possibility to realize stable contacts, in terms of Ge-segregation and strain relaxation.

In this work, in order to investigate further the potentiality of Zr/SiGe(C) contacts, we have focused on its electrical properties, after annealing in a RTA furnace at a temperature ranging from 600 to 800°C for 5 min. We will report changes in the Schottky Barrier Height (SBH) on n-type and p-type SiGeC layers due to different annealing conditions.

2. Experimental procedures

The samples used consisted of p-type (B doped) and n-type (P doped) relaxed and pseudomorphic $Si_{1-x-y}Ge_xC_y$ layers ($x = 0.10\text{–}0.17$ and $y = 0.005\text{–}0.013$). The 100 nm-SiGeC layers were grown by rapid thermal chemical vapor deposition on a (100) oriented Si substrate of the same type at 550–570°C. Prior to metal deposition, each SiGe sample was cleaned using the same procedure: a standard chemical degreasing was followed by a dip in diluted HF for 30 s and a final rinsing in deionised water. 16–80 nm-Zr films were deposited in a dc-magnetron sputtering chamber. Photolithography followed by a photoresist annealing at 150°C and a selective etch were used to define 96 diodes of different areas ($0.071\text{–}0.384$ mm^2). Heat treatments only concerned the 16 nm-Zr film and were performed in a RTA system under Ar/H_2 atmosphere at a temperature from 600 to 800°C for 5 min. The thickness of Zr films was chosen so that the Zr/SiGe reaction would not completely consume the SiGe layer.

During annealing, the sample was maintained between two other samples of Zr/Si. In this case, the sample surface was protected and was not directly in contact with the furnace atmosphere. This procedure limits oxidation in an acceptable range. The Schottky barrier height (SBH) values Φ_b were determined by the $I\text{–}V$ measurement of the reverse current at room temperature. To improve the resolution, at least 30 diodes were systematically characterized in order to obtain a mean SBH value with good accuracy and informations on the homogeneity of the interface. More details of this procedure can be found in Ref. [14].

3. Results and discussion

Fig. 1 exhibits the SBH as a function of the annealing temperature for both $Zr/p\text{-}Si_{0.83}Ge_{0.17}$ and $Zr/n\text{-}Si_{0.84}Ge_{0.16}$ contacts. For p-type, the RTA treatment mostly leads to a decrease of the SBH. For n-type, the SBH first increases up to 600°C and then decreases. After a reaction at 800°C, Φ_{bp} and Φ_{bn} always exhibit lower values than those before annealing. In principle, according to the theory of barrier formation developed by Schottky, the sum $\Phi_{bp} + \Phi_{bn}$ is expected to give the band-gap of the semiconductor E_g [15]. Indeed, we have shown that this model describes rather well our results on as-deposited samples [14]. After RTA at 800°C, the slight strain relaxation of the SiGe underlayer should lead to an increase of E_g and cannot explain the decrease of both Φ_{bp} and Φ_{bn}. Therefore, the SBH lowerings are likely due to the formation of zirconium germanosilicide and interface defects even if no Ge-segregation was evidenced by TEM. In order to better understand our results, we have analyzed the SBH values, for samples as-deposited and annealed at 800°C.

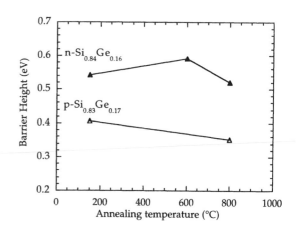

Fig. 1. Schottky barrier height values as a function of the annealing temperature for $Zr/p\text{-}Si_{0.83}Ge_{0.17}$ and $Zr/n\text{-}Si_{0.84}Ge_{0.16}$ contacts.

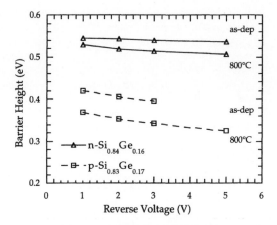

Fig. 2. Variations of the Schottky barrier heights as a function of the reverse voltage for $Zr/p-Si_{0.83}Ge_{0.17}$ and $Zr/n-Si_{0.84}Ge_{0.16}$ contacts.

The variations of SBHs as a function of the reverse voltage have been studied for both $Zr/p-Si_{0.83}Ge_{0.17}$ and $Zr/n-Si_{0.84}Ge_{0.16}$ contacts (Fig. 2). The decrease of SBH with voltage, before and after annealing, is well described by the thermionic emission, taking into account the same image force lowering and field effect through an interfacial layer. This shows that the conduction mode is conserved and that the degradation of the interface, if it exists, is limited.

Fig. 3 shows typical SBH histograms of $Zr/n-Si_{0.84}Ge_{0.16}$ diodes obtained from reverse current measured at 2 V on samples as-deposited and annealed at 800°C. The standard deviation of SBH increases with the temperature from 1.5 to 7.5 meV. The soft scattering in SBH, before annealing, accounts for a very homogeneous metal/semiconductor (M/SC) interface on the whole surface of the sample. The slight increase of the standard deviation, after thermal treatment, suggests that some degradation has occurred at the interface M/SC and may show some inhomogeneities at the interface. Nevertheless, it is noteworthy that the value of the standard deviation after annealing remains small. We have also investigated the influence of the annealing on the $Zr/p-Si_{0.83}Ge_{0.17}$ contact. An increase of the standard deviation from 1.9 to 4.1 meV is observed. The same trends with annealing are observed for p-type and n-type and may show a slight reduction of the quality of the interface.

4. Conclusions

In conclusion, the $C49-Zr(Si_{1-z}Ge_z)_2$ compound is formed after severe annealing conditions (800°C, 5 min) of $Zr/Si_{1-x-y}Ge_xC_y$ contact and no Ge-segregation is observed. The reaction only leads to a slight strain relaxation in the SiGe epilayer. The presence of

carbon still improves the stability of the system and no relaxation occurs in the unreacted layer. We have also shown that for Zr contacts on both p-type and n-type SiGe(C) layers, the thermionic mode is preserved after annealing. We have observed that whatever the type there is a decrease of the SBH after the annealing at 800°C and a slight increase of the SBH standard deviation. These last results suggest that few interface defects are created during annealing. To understand

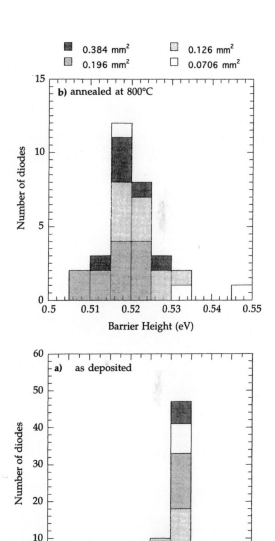

Fig. 3. Histograms of Schottky barrier heights of $Zr/n-Si_{0.84}Ge_{0.16}$ with areas varying from 0.071 to 0.384 mm^2: (a) as-deposited and (b) after annealing at 800°C. The histograms are obtained from reverse current measured at −2 V.

the decrease of the sum $\Phi_{bp} + \Phi_{bn}$, further investigations are required.

References

[1] Osten HJ, Bugiel E, Zaumseil P. Appl Phys Lett 1994;64:3440.

[2] Powell AR, Iyer SS. Jpn J Appl Phys 1994;33:2388.

[3] Williams RL, Aers GC, Rowell NL, Brunner K, Winter W, Eberl K. Appl Phys Lett 1998;72:1320.

[4] Aldrich DB, Chen YL, Sayers DE, Nemanich RJ, Ashburn SP, Öztürk M. J Appl Phys 1995;77:5107.

[5] Eyal A, Brener R, Beserman R, Eizenberg M, Atzmon Z, Smith DJ, Mayer JW. Appl Phys Lett 1996;69:64.

[6] Glück M, Schüppen A, Rösler M, Heinrich W, Hersener J, König U, Yam O, Cytermann C, Eizenberg M. Thin Solid Films 1995;270:549.

[7] Buxbaum A, Eizenberg M, Raizman A, Schäffler F. Appl Phys Lett 1991;59:665.

[8] Aubry V, Meyer F, Laval R, Clerc C, Warren P, Dutartre D. Mater Res Soc Symp Proc 1994;320:299.

[9] Ridgway MC, Rao MR, Baribeau J-M. Mater Res Soc Symp Proc 1994;320:329.

[10] Nur O, Willander M, Hultman L, Radamson HH, Hansson GV, Sardela MR, Greene JE. J Appl Phys 1995;78:7063.

[11] Wang Z, Aldrich DB, Nemanich RJ, Sayers DE. J Appl Phys 1997;82:2342.

[12] Aubry-Fortuna V, Barthula M, Perrossier J-L, Meyer F, Demuth V, Strunk HP, Chaix-Pluchery O. J Vac Sci Technol B 1998;B16:1659.

[13] Aubry-Fortuna V, Eyal A, Chaix-Pluchery O, Barthula M, Meyer F, Eizenberg M. Appl Phys Lett 1998;73:1248.

[14] Meyer F, Mamor M, Aubry-Fortuna V, Warren P, Bodnar S, Dutartre D, Regolini JL. J Electron Mater 1996;25:1748.

[15] Sze SM. Physics of semiconductor devices. 2nd ed. New York: Wiley-Interscience, 1981.

PERGAMON

Materials Science in Semiconductor Processing 1 (1998) 267–270

MATERIALS
SCIENCE IN
SEMICONDUCTOR
PROCESSING

Ultra high temperature rapid thermal annealing of GaN

X.A. Cao [a,*], R.K. Singh [a], S.J. Pearton [a], M. Fu [b], J.A. Sekhar [b],
R.G. Wilson [c], J.C. Zolper [d], J. Han [e], D.J. Rieger [e], R.J. Shul [e]

[a] *Department of Materials Science and Engineering, University of Florida, Gainesville, FL 32611, USA*
[b] *Micropyretics Heaters International, Inc. Cincinnati, OH 45212, USA*
[c] *Consultant, Stevenson Ranch, CA 91381, USA*
[d] *Office of Naval Research, Arlington, VA 22217, USA*
[e] *Sandia National Laboratories, Albuquerque, NM 87185, USA*

Abstract

All of the major acceptor (Mg, C, Be) and donor (Si, S, Se and Te) dopants have been implanted into GaN films grown on Al_2O_3 substrates. Annealing was performed at 1100–1500°C, using AlN encapsulation. Activation percentages of $\geq 90\%$ were obtained for Si^+ implantation annealed at 1400°C, while higher temperatures led to a decrease in both carrier concentration and electron mobility. No measurable redistribution of any of the implanted dopants was observed at 1450°C. © 1999 Elsevier Science Ltd. All rights reserved.

1. Introduction

The use of selective area implantation to create channel and/or contact regions is the basis of standard metal semiconductor field effect transistor (MESFET) technology in GaAs, and a similar process is desirable for GaN electronics [1, 2]. Currently, most GaN-based electronic devices for high power, high frequency, high temperature applications are heterostructure FETs [3–7], and for these devices implantation is also useful for increasing doping in the source/drain regions for improved contact resistance [8].

Past work has shown that anneal temperatures of 1100–1150°C produce reasonably good activation efficiencies for Si^+ or Mg^+ implant doses up to $\sim 10^{14}$ cm^{-2} [8–11]. However considerable lattice damage remains for higher dose implants annealed at these temperatures, producing a clear need for furnaces capable of 1400–1500°C [11–16]. Conventional rapid thermal processing(RTP) systems generally reach a maximum temperature of ~ 1300°C. It is desirable that the time at elevated temperature be minimized because of the high vapor pressure of N_2 above GaN and the need to prevent dissociation of the surface [17, 18]. Several different surface protection schemes have been reported for high temperature annealing of GaN, including provision of NH_3 ambients, high pressure N_2 ambients, AlN encapsulation or use of granulated InN or GaN powder within the reservoirs of a graphite susceptor in which the implanted sample is contained [19, 20]. In terms of utility in a fabrication line, the AlN cap approach seems the most effective. The AlN can be deposited by reactive sputtering and selectively removed after the annealing processing by KOH etching [21].

In this paper we describe the use of a novel high temperature RTP system for annealing of implanted GaN at temperatures up to 1500°C. Extremely good activation efficiencies for Si^+ implants were obtained ($\geq 90\%$), while little redistribution was observed for all the common donors (Si, S, Se, Te) and acceptor (Mg, C, Be) species.

* Corresponding author.

2. Experimental

Epitaxial GaN layers 2–3 μm thick were grown on c-axis Al_2O_3 substrates at ~1040°C by metal organic chemical vapor deposition. The layers were nominally undoped ($n = 1$–8×10^{16} cm^{-3}). Implantation of Si$^+$, S$^+$, Se$^+$, Te$^+$, Be$^+$, C$^+$ or Mg$^+$ ions was performed at 25°C, at doses of 1–5×10^{15} cm^{-2} and energies designed to place the projected range at 1200–1500 Å. The samples were deposited with 1000 Å of AlN by reactive sputtering.

Annealing was performed in an MHI Zapper RTP furnace, which employs molybdenum intermetallic composite heating elements. These are maintained at constant temperature with the sample inserted and removed via a motor-driven actuator to achieve high ramp-up and ramp-down rates. Annealing was performed at 1100–1500°C for dwell times of 10 s. Typical time–temperature profiles for 1400 and 1500°C are shown in Fig. 1. After annealing the AlN was removed by etching in 0.1 M KOH solution at 70°C. The electrical properties of the Si$^+$ implanted samples were obtained from Van der Pauw geometry Hall measurements using HgIn contacts alloyed at 420°C for 3 min. Redistribution of the implanted species was

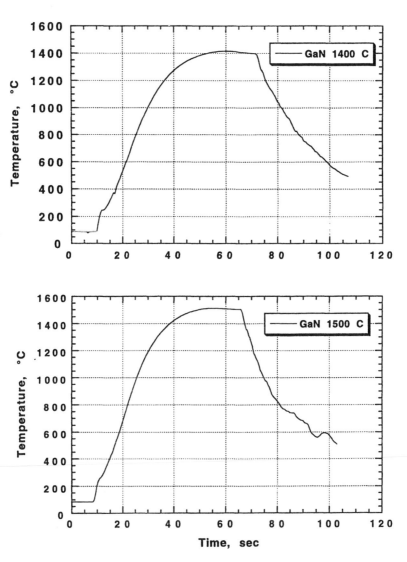

Fig. 1. Time–temperature profiles for 10 s anneals at both 1400 and 1500°C.

examined by performing secondary ion mass spectrometry (SIMS) using a Cameca system.

3. Results and discussion

Fig. 2 shows an Arrhenius plot of sheet carrier concentration in Si^+ implanted material. In the AlN encapsulated samples activation occurs with an activation energy of ~5.2 eV before saturating at ~1400°C. We interpret this activation energy as the average required to move the interstitial Si atom to a vacant substitutional site by short-range diffusion and to simultaneously remove compensating point defects so that the Si is electrically active. Note that at 1500°C the sheet electron density decreases, and this was accompanied by a decrease in carrier mobility. This increase in compensation is consistent with Si beginning to occupy both Ga (where it is a donor) and N sites (where it is an acceptor). This is commonly observed with Si implantation in other III–V materials [22]. The peak n-type doping level we obtained is ~5×10^{20} cm^{-3}. This high carrier concentration enhances emission over the barrier on metal contacts deposited on the material. For example, for both W and WSi$_x$ sputter deposited on Si-implanted material activated by annealing, we obtained specific contact resistances of ~10^{-6} $\Omega \cdot$ cm^2 after annealing in the range 600–900°C. This combination is particularly effective for producing high quality, stable ohmic contacts on GaN, since there is no measurable reaction of the W or WSi$_x$ with the semiconductor at 900°C. This

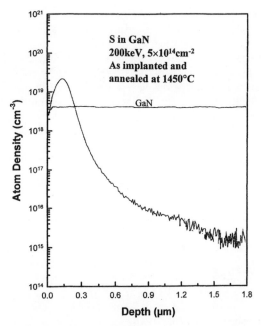

Fig. 3. SIMS profiles of 200 keV S$^+$ implants in GaN before and after annealing at 1450°C for 10 s.

is superior to the more commonly employed Ti/Al and Ni/Au metallizations on GaN.

Si$^+$ implantation produces the best n-type doping of GaN, but there is also interest in the group VI donors: S, Se and Te. From preliminary measurements, we obtained ≤40% activation for these dopants under the

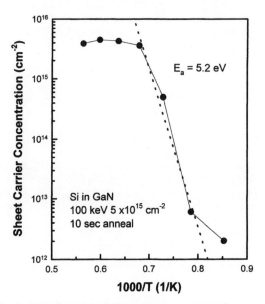

Fig. 2. Arrhenius plot of sheet electron concentration versus inverse anneal temperature for Si$^+$ implanted GaN.

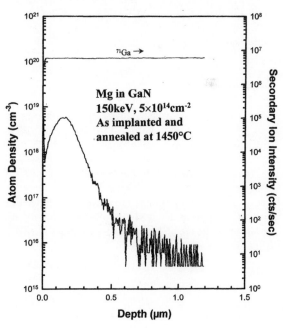

Fig. 4. SIMS profiles of 150 keV Mg$^+$ implants in GaN before and after annealing at 1450°C for 10 s.

same conditions that produced 90% activation with Si. There was basically no redistribution of any of the dopants to 1450°C: given the resolution of the SIMS measurements, this indicates that the diffusivity for each of these elements at 1450°C is $\leq 10^{-13}$ cm^2 > s^{-1}. These are clearly the lowest diffusivities of these elements in any compound semiconductor, and emphasize the stable nature of implanted GaN devices. Fig. 3 shows the data for S$^+$ implants. In a similar fashion, the atomic profiles of Mg, C and Be were measured before and after annealing at 1450°C, there was no detectable redistribution for Mg and C. Fig. 4 shows the data for Mg$^+$ implants. Thus, the diffusivities of these acceptor species are also $\leq 10^{-13}$ cm$^2 \cdot$ s^{-1} at 1450°C. For Be, a slight amount of diffusion was observed at 900°C ($\sim$300 Å at full-width–half-maximum), but no motion at higher temperatures. This is consistent with defect-assisted motion of the Be, which ceases once the implant damage is annealed.

4. Summary and conclusions

The most common acceptor and donor dopants have been implanted into GaN, and annealed at 1100–1500°C for dwell times of 10 s at the peak temperatures. No measurable redistribution was observed for any of the implanted species at 1450°C. AlN has proven to be an effective encapsulant at these high temperatures and can be selectively removed in KOH solutions, there was no evidence of interdiffusion between the AlN and GaN. Annealing at 1400°C produced the highest activation for implanted Si ($\sim$90%), while higher temperatures led to an increase in self compensation.

Acknowledgements

The work of UF is partially supported by grants from NSF-DMR (L. Hess), DARPA/EPRI (E.R. Brown/J. Melcher) and a subcontract from MHI, who are supported by a BMDO SBIR grant (F19628-97-C-0092) monitored by Dr. Joe Lorenzo. RGW is partially supported by a grant from ARO (Dr. J.M. Zavada). Sandia is a multiprogram laboratory, operated by Sandia Corporation, a Lockheed–Martin company, for the US Department of Energy under contract No. DE-AC04-94 AL 85000.

References

[1] Zolper, J.C. In: Pearton, S.J., editor. GaN and related materials. New York: Gordon and Breach, 1997.

[2] Zolper JC, Shul RJ. MRS Bull 1997;22:36.

[3] Binari SC, Rowland LB, Kruppa W, Kelner G, Doverspike K, Gaskill DK. Electron Lett 1994;30:1248.

[4] Burm J, Chu K, Davis WA, Schaff WJ, Eastman LF, Eustis TJ. Appl Phys Lett 1997;70:464.

[5] Chen Q, Khan MA, Yang JW, Sun CJ, Shur MS, Park H. Appl Phys Lett 1996;69:794.

[6] Wu Y-F, Keller BP, Keller S, Kapolnek D, Kozodoy P, DenBaars SP, Mishra UK. Appl Phys Lett 1996;69:1438.

[7] Nguyen NX, Keller BP, Keller S, Wu YF, Le M, Nguyen C, DenBaars SP, Mishra UK, Grider D. Electron Lett 1997;33:334.

[8] Zolper JC, Shul RJ, Baca AG, Wilson RG, Pearton SJ, Stall RA. Appl Phys Lett 1996;68:2273.

[9] Zolper JC, Crawford MH, Pearton SJ, Abernathy CR, Vartuli CB, Yuan C, Stall RA. J Electron Mater 1996;25:839.

[10] Pearton SJ, Abernathy CR, Vartuli CB, Zolper JC, Yuan C, Stall RA. Appl Phys Lett 1995;67:1435.

[11] Zolper JC, Wilson RG, Pearton SJ, Stall RA. Appl Phys Lett 1996;68:1945.

[12] Tan HH, Williams JS, Zou J, Cockayne DJH. Pearton SJ, Stall RA. Appl Phys Lett 1996;69:2364.

[13] Zolper JC, Crawford MH, Tan HH, Williams JS, Zou J, Cockayne DJH, Pearton SJ, Karlicek RF. Appl Phys Lett 1997;70:2729.

[14] Strite S, Epperlein PW, Dommen A, Rockett A, Broom RF. Mat Res Soc Symp Proc 1996;395:795.

[15] Parikh N, Suvkhanov A, Lioubtchenko M, Carlson E, Bremser M, Bray D, Davis R, Hunn J. Nucl Instr Meth B 1997;127/128:463.

[16] Zolper JC, Han J, Biefeld RM, Van Deusen SB, Wampler WR, Reiger DJ, Pearton SJ, Williams JS, Tan HH, Stall R. J Electron Mater 1998;27:179.

[17] Porowski S., Grzegory I. In: Pearton, S.J., editor. GaN and related materials. New York: Gordon and Breach, 1997.

[18] Newman N. In: Pankove, J., Moustakas, T.D. GaN, vol. I. New York: Academic Press, 1998.

[19] Hong J, Lee JW, MacKenzie JD, Donovan SM, Abernathy CR, Pearton SJ, Zolper JC. Semicon Sci Technol 1997;12:1310.

[20] Zolper JC, Reiger DJ, Baca AG, Pearton SJ, Lee JW, Stall RA. Appl Phys Lett 1996;69:538.

[21] Mileham JR, Pearton SJ, Abernathy CR, MacKenzie JD, Shul RJ, Kilcoyne SP. Appl Phys Lett 1995;67:1119.

[22] Pearton SJ, Williams JS, Short KT, Johnson ST, Jacobsen DC, Poate JM, Gibson JM, Boerma DO. J Appl Phys 1989;65:1089.

PERGAMON

Materials Science in Semiconductor Processing 1 (1998) 271–274

MATERIALS
SCIENCE IN
SEMICONDUCTOR
PROCESSING

Influence of vapor phase pre-oxide-cleaning on the oxidation characteristics

B. Froeschle *, N. Sacher, F. Glowacki, T. Theiler

STEAG AST Elektronik GmbH, Daimlerstrasse 10, 89160 Dornstadt, Germany

Abstract

In this paper the oxide growth with and without annealing in nitric oxide (NO) is investigated for different types of pre-oxidation cleaning.

Two types of vapor phase cleaning (VPC), so called VPC1 and VPC2, are investigated and compared to standard wet cleaning. The VPC1 cleaning uses anhydrous HF (AHF)/methanol for the removal of the native oxide. The VPC2 cleaning is similar to the VPC1 with an additional ozone cleaning step after the native oxide removal. The oxidation rate in a RTO module, which is integrated in the same cluster tool increases for integrated pre-oxidation cleaning (VPC) compared to the standard wet cleaning. These results correlate to the increase in fluorine content in the oxide using the VPC pre-oxidation cleaning. The NO annealed oxides show a more significant increase in oxidation rate using the VPC2 cleaning type, which has an ozone step after the native oxide removal. This seems also to be correlated to the difference in fluorine content. © 1999 Published by Elsevier Science Ltd. All rights reserved.

1. Introduction

As the film thickness decreases, Si/SiO_2 interface properties play a more significant role with respect to gate dielectric processing and integrity. Thus, pre-gate surface preparation becomes one of the most critical steps in future device technologies.

Integrated single wafer manufacturing in cluster tools is expected to be important for 0.18 μm technology and beyond [1–3]. Due to the fact that standard wet cleaning can not be easily integrated into a cluster tool, vapor phase cleaning using anhydrous HF (AHF)/methanol for native oxide removal becomes more and more important.

NO annealed thin oxides have gained a lot of interest due to the superior electrical characteristics and the efficient boron diffusion barrier properties [4]. Therefore, the influence of the different cleaning procedures on the oxidation followed by a NO annealing needs further attention.

2. Experimental

2.1. Cluster tool

The experiments are performed in a ASMI/STEAG AST Hot Cluster Advance 800 Polygon. This single wafer 200 mm rapid thermal processing cluster tool integrates vapor phase cleaning (VPC), rapid thermal oxidation/nitridation (RTO/RTO-N), nitride deposition (CVD) and in-situ doped polysilicon deposition (CVD).

2.2. Sample preparation

2.2.1. Vapor phase cleaning
Two different types of vapor phase cleaning, so called VPC1 and VPC2, are compared to standard wet cleaning.

1. VPC1 cleaning: the VPC1 type of cleaning can be divided into three main steps.

- Step 1: wetting of the wafer surface before the native oxide removal using methanol supplied via a

---

* Corresponding author.

nitrogen bubbler. The wafer temperature is held at 40°C and the chamber pressure is 400 mbar.

- Step 2: removal of the native oxide at a temperature of 40°C and a pressure of 100 mbar using a mixture of anhydrous HF (AHF) and nitrogen.
- Step 3: desorption of the etch products under reduced pressure.

2. VPC2 cleaning: to improve the cleaning sequence of the VPC1 cleaning, for the VPC2 cleaning an additional step is performed by keeping the wafer in the VPC module and introducing ozone at 200°C and a pressure of 4 mbar for 60 s after the desorption step of the VPC1 cleaning. The ozone concentration is 1% in oxygen. During this ozone step the wafer is illuminated with ultra violet (UV) light.

2.2.2. RTO/RTO-N

The oxidation (RTO) is carried out directly after the cleaning in the RTO module at a pressure of 500 mbar and in the temperature range 950 to 1130°C.

Nitrided oxides (RTO-N) are produced by annealing the oxides in 100% NO at the same temperature and at the same pressure than the oxidation. The NO annealing is performed after a gas exchange step under reduced pressure and a nitrogen purge at a temperature of 750°C.

3. Results

3.1. RTO (pure oxidation)

The dependence of the oxide thickness on the oxidation time is shown in Fig. 1. The oxide grown after the VPC2 cleaning sequence is about 0.4 nm thicker than the oxide obtained after the VPC1 cleaning sequence. The oxidation characteristics for both type of pre-oxidation cleaning follow the Deal–Grove model [5] for silicon dioxide growth.

In Fig. 2 the temperature dependence of the oxide thickness is shown. The values are obtained by plotting the oxide thickness difference between VPC pre-cleaned and standard cleaned samples against the temperature. For low temperatures (950°C) the difference in oxidation rate is small and can be neglected. For temperatures higher than 1000°C the difference obtained in the case of the VPC1 cleaning is 0.1 nm whereas the difference obtained in the case of the VPC2 cleaning is 0.4 nm.

To understand this behavior, the different types of cleaning have been analyzed more in detail.

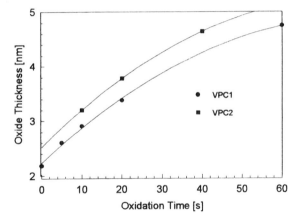

Fig. 1. Oxide thickness obtained by RTO at 1000°C after two different types of vapor phase cleaning (VPC1 and VPC2) as a function of the oxidation time.

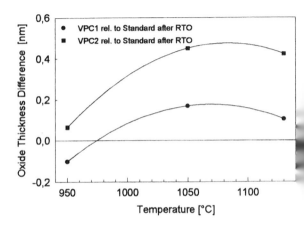

Fig. 2. Difference in oxide thickness obtained by RTO after VPC1 and VPC2 cleanings compared to standard wet cleaning.

3.1.1. VPC1 cleaning

After the VPC1 cleaning, the native oxide is completely removed from the silicon surface. This can be shown by performing contact angle measurements as shown previously [6]. The contamination of the silicon surface is measured by X-ray photoelectron spectroscopy (XPS) [6]. No native oxide is grown before the oxidation in the RTO module due to the transport of the wafer in the cluster tool under controlled ambient. Fluorine has been measured to be about 0.2 monolayer (ML). After performing a rapid thermal oxidation of a 4 nm oxide at a temperature of 1000°C and a pressure of 500 mbar, the fluorine content was measured again by XPS and found to be less than 0.1 ML [7].

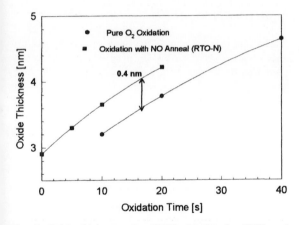

Fig. 3. Oxide thickness after VPC2 cleaning for RTO and RTO with NO anneal (RTO-N) samples performed at 1000°C and 500 mbar as a function of the oxidation time.

3.1.2. VPC2 cleaning

During the VPC2 cleaning step a thin oxide layer of about 2 ML is grown on the AHF/methanol cleaned silicon surface in the VPC module, because of the oxidation behavior of ozone in the subsequent ozone cleaning step. The fluorine content of this thin oxide is 0.82 ML has been measured by XPS. During the oxidation in the RTO module, this thin oxide acts as a source of fluorine, which diffuses through the grown oxide layer and piles up at the surface of the RTO grown oxide. After a 4 nm oxide growth at 1000°C and 500 mbar in the RTO module, the fluorine content is lowered to about 0.3 ML [7].

3.2. RTO-N (pure oxidation followed by a NO anneal)

The oxide thickness after a NO anneal increases compared to non-annealed samples.

For example, after the VPC2 cleaning, the increase in oxide thickness of a 1000°C oxide followed by a 30 s NO anneal at the same temperature is about 0.4 nm as shown in Fig. 3.

Fig. 4 shows the temperature dependence of the thickness differences obtained for RTO and NO annealed wafers using VPC1 and VPC2 cleaning, respectively. The effect of the VPC1 cleaning on the oxidation growth rate with a NO anneal is comparable to the oxidation without such annealing (Fig. 3) and can therefore be neglected. In the case of the VPC2 cleaning, the obtained thickness difference is more important even at low temperatures. At temperatures higher than 1000°C, the difference in oxide thickness is about 0.6 nm, which is much higher than expected from the RTO samples without NO anneal. Therefore an additional effect is taking place for the oxidation followed by a NO anneal.

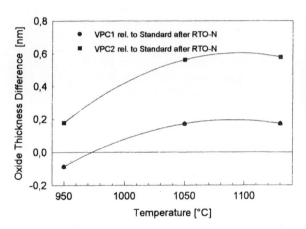

Fig. 4. Difference in oxide thickness obtained by oxidation followed by NO anneal (RTO-N) after two types of VPC cleanings, so called VPC1 and VPC2 cleaning, compared to standard wet cleaning.

3.2.1. Investigation of nitrogen and fluorine distribution in RTO-N samples

The nitrogen and the fluorine distributions in NO annealed samples at 1050°C are analyzed by secondary ion mass spectroscopy (SIMS) using Cs^+ bombardment and Cs(N) and Cs(F) detection.

For both standard wet cleaning and VPC1 cleaning the nitrogen distribution is found to be similar, however, a difference in fluorine concentration is observed. The VPC1 cleaned samples have a slightly higher fluorine content compared to the standard wet cleaned samples. This is attributed to the use of AHF in the VPC1 cleaning sequence and is in accordance to the XPS measurements performed on pure RTO-oxides.

The VPC1 cleaning sequence, especially at low temperatures, has no effect on the oxide grown during the NO annealing. However, in the case of the VPC2 cleaning sequence, which includes an ozone treatment, a remarkable increase in oxide thickness is shown for the RTO samples with NO anneal. This higher oxidation rate for VPC cleaned samples can be attributed to the much higher fluorine content due to this cleaning sequence., This is confirmed by some previous studies [7].

3.3. Discussion

The mechanism of fluorine incorporation by using vapor phase pre-oxidation cleaning seems to be comparable to the addition of NF_3 gas during the oxidation process as described by Joshi et al. [8].

During the first initial growth, a fluorine rich silicon oxide layer is formed, where fluorine can saturate the dangling bonds at the silicon/oxide interface. This network leads to strain relaxation at the interface.

The mechanism of the enhanced oxidation rate is not fully understood. It could be correlated to:

- the strain relaxation [8] as cited above.
- the presence of a less dense layer on the top of the oxide film, which does not act as a diffusion barrier for oxygen.
- or a mechanism of enhanced diffusion of oxygen due to the presence of fluorine [9].

4. Conclusion

The results shown in this paper give some ideas about the influence of vapor phase cleaning on the oxide growth of pure RTO and RTO followed by NO anneal (RTO-N) samples. The results are compared to former studies dealing with the surface analysis of vapor phase cleaned wafers by X-ray photoelectron spectroscopy (XPS) [6].

It is found, that the oxidation rate depends on the different wafer surface cleaning procedures. Comparing the VPC1 and VPC2 cleanings, the variation in oxide thickness can be attributed to the difference in fluorine content after the pre-oxidation cleaning sequence. This observation is in good agreement to other publications [8, 10, 11]. The mechanism is not yet fully understood. Further investigations have to be performed to explain this mechanism of this enhanced oxide growth rate.

Acknowledgements

The authors would like to thank A. Gschwandtner and G. Innertsberger from Siemens AG in Munich for their support in this work and are grateful for the analytical supports of I. Kasko and M. Lucassen from the Fraunhofer Institute for Integrated Circuits in Erlangen.

References

[1] Cleaning and contamination monitoring systems for the semiconductor industry. San Jose, CA: VLSI Research, April 1986.
[2] Ruzyllo J. Microcontamination 1988;6(3):39.
[3] Khilnami A. Cleaning semiconductor surfaces: facts and foibles. In: Mittal KL, editor. Particles on surfaces 1. Detection, adhesion and removal. New York: Plenum Press, 1986. p. 17.
[4] Bhat M, Han LK, Wristers D, Kwong DL, Fulford J. Appl Phys Lett 1995;66(10):1225.
[5] Deal BE, Grove AS. J Appl Phys 1965;36:3770.
[6] Froeschle B, Glowacki F, Kasko I, Oechsner R, Bauer A, Schneider C. Proceedings of the 192nd Electrochemical Society Meeting, vol. 97-35. 1997. p. 415.
[7] Froeschle B, Theiler T. Abschlussbericht ADICT. 1998.
[8] Joshi AB, Lo GQ, Ahn J, Ting W, Kwong DL. SPIE 1991;1595:177–94.
[9] Morita M, Kubo T, Ishihara T, Hirose M. Appl Phys Lett 1990;56:2255.
[10] Wei L, Yuan-sen Z, Yang-shu Z. Proceedings of the Symposium on the Physics and Chemistry of SiO and SiO$_2$ Interface. Plenum Press, 1988. p. 103–10.
[11] Lo GQ, Ting W, Kwong D-L, Kuehne J, Magee CW. IEEE Electron Device Lett 1990;11(11):511–3.

PERGAMON

MATERIALS
SCIENCE IN
SEMICONDUCTOR
PROCESSING

Materials Science in Semiconductor Processing 1 (1998) 275–279

Rapid thermal oxidation of highly in situ phosphorus doped polysilicon thin films

S. Kallel*, B. Semmache, M. Lemiti, A. Laugier

Laboratoire de Physique de la Matière (UMR-CNRS 5511), Bât 502, Institut National des Sciences Appliquées de Lyon, 20 Avenue Albert Einstein, 69621 Villeurbanne Cedex, France

Abstract

Thin polycrystalline silicon oxide (polyoxide) layers were obtained by rapid thermal oxidation (RTO) growth in a cold wall rapid thermal processing (RTP) reactor using a pure dry O_2 (99.998%) atmospheric pressure. Oxidations were carried out both on classical $POCl_3$-doped low pressure chemical vapor deposition (LPCVD) polysilicon films and in situ PH_3-doped RTLPCVD polysilicon layers in order to point out the in situ integrated multiprocessing offered by RTP treatments. RTO films thicknesses were obtained both by means of spectrometric ellipsometry (SE) and capacitance–voltage (C–V) measurements. It has been noted that polyoxide growth rates are more accelerated when the oxidation temperature increases and even more accelerated when the phosphorus dopant concentration is greater than 10^{20} at/cm^3. In addition, X-ray photoelectron spectroscopy (XPS) characterization revealed the presence of a Si-rich polyoxide (SiO_x, $1 < x < 2$) layer at the vicinity of the polysilicon/polyoxide interface. However, sheet resistivity measurements and secondary ion mass spectrometry (SIMS) analysis showed that the RTO polyoxide process plays an effective role both as dopant activation step and a cap layer formation to block phosphorus outdiffusion. This process scheme may permit a tight control of the poly/polyoxide interface and thus insure a good device performance and reproducibility. © 1999 Elsevier Science Ltd. All rights reserved.

1. Introduction

High quality thin oxide on polycrystalline silicon (poly-Si) films is a desirable material for submicron metal/oxide/semiconductor (MOS), bipolar complementary MOS (BiCMOS) and memory cells (dynamic random access memory, erasable programmable read only memory) very large scale integration (VLSI) devices. Indeed, oxide layers grown on poly-Si seem to display some reliability problems, as compared with those grown on planar crystalline silicon (c-Si) substrates. Many investigations have been devoted to the polyoxide growth process [1–5]. It appears mainly that the poly-Si structural texture induces oxidation rate anisotropy because the oxide film thickness is differently dependent on the grain orientation. In addition, intergranular oxidation of poly-Si is responsible for surface roughness. Some authors suggested that high oxidation temperatures can lead to a smoother interface due to stress relief by viscous flow in the growing oxide layer [6, 7]. Otherwise, higher oxidation temperatures tend to result in a more diffusion-limited growth which presents a more uniform growth interface resulting in fewer asperities [8]. However, increasing the oxidation temperature is not compatible with the need to reduce the thermal budget. Consequently, rapid thermal oxidation (RTO) seems to be an alternative technique to conventional furnace oxidation (CFO) insofar as the overall thermal budget is greatly reduced. Few works concerning RTO polyoxide growth have been cited in the literature [9, 10] comparing to RTO oxide growth which was extensively studied and successfully used in microelectronics [11] and photovoltaics technologies [12, 13]. Oxidation of poly-Si can be important to the performance of poly-Si emitter solar cells and especially for small-grain size poly-Si films. Indeed, because of the enhanced dopant diffusion at

---

* Corresponding author.

1369-8001/99/$ - see front matter © 1999 Elsevier Science Ltd. All rights reserved.
PII: S 1 3 6 9 - 8 0 0 1 (9 8) 0 0 0 3 8 - 9

the grain boundaries (GB), carrier recombination within the grains is possible. Since the GB display a higher oxidation rate, polyoxide could be used to passivate these regions and then improve the current transport toward the emitter junction.

In the present work, the RTO process is used in order to activate phosphorus dopants and also to grow a surface passivation polyoxide layer on an in situ PH_3-doped RTLPCVD poly-Si thin film emitter [14]. Indeed, the RTO cycle is performed sequentially after a poly-Si deposition step in the same lamp-heated reactor, without handling of the wafer. We have investigated RTO polyoxide growth and particularly effects of oxidation temperature and high phosphorus doping level on oxidation kinetics.

2. Experimental

Two sets of P-doped polysilicon films were used. LPCVD polysilicon films (series c) with a thickness of 200 nm were deposited by SiH_4 (silane) thermal decomposition at standard conditions (620°C, 300 mTorr). The starting material was 2 inches in diameter (100) oriented Cz-grown p-type Si (0.5–1 Ω cm) wafer covered with a 200 nm thick thermal oxide. Phosphorus doping was made by $POCl_3$ source deposition and drive-in at 1025°C which yields a dopant concentration of 10^{21} at/cm³ measured by SIMS analysis. The phosphorus silicon glass (PSG) layer grown during the predeposition step was buffered-HF stripped. The second set (series a) of samples consisted of 200 nm thick RTLPCVD in situ PH_3-doped poly-Si films prepared also by silane decomposition at deposition and a total pressure of 750°C and 2 mbar, respectively. Under these conditions, totally crystallized poly-Si films with a chemical phosphorus concentration of 1.2×10^{20} at/cm³ are obtained. Other details on RTLPCVD deposition experimental procedure are given elsewhere [14]. RTO processes were carried out in a FAV4 (Jipelec™) cold wall reactor. Before the oxidation process is started, the reactor is pumped down during 1 min, afterwards a preheating step of the reactant gas at 700°C is first applied in order to adjust the gas flow rate (1000 sccm), then the ramp up step (100°C/s) starts toward the growth plateau of variable duration (5 to 100 s) at oxidation temperatures of 1000 and 1100°C, then cooling under a reducing ambient ($Ar/H_2 = 10\%$), in order to flash the reaction chamber and stop the oxidation reaction, takes place. Polyoxide thicknesses were determined either by an SE (series a) or $C–V$ technique (series c) at high HF frequency using a non-destructive mercury (Hg)-probe contact assuming a dielectric relative permittivity of 3.9. SE analysis was carried out in a two-step procedure. The first step was the determination of

an equivalent substrate. This consists of measuring poly-Si films optical constants in order to find the reference structure. Because oscillating interferences of the multilayer structure are observed at high wavelengths, measurements are limited to the short wavelength region (240–420 nm) with a step width of 5 nm. The reference substrate is considered to be absorbed in the UV wavelength region. In the second step, the method proposed in Ref. [15] is used to determine the thickness of the top polyoxide assumed as a transparent layer.

3. Results and discussion

3.1. Oxide growth kinetics

In Fig. 1 RTO growth kinetics of oxides are plotted both on virgin standard Cz-Si substrates and poly-Si films at different temperatures. Data of samples (b) and (d) were taken from a previous study [16]. It can be noted that the growth kinetics are more accelerated as the oxidation temperature increases and much more when phosphorus dopant concentration is greater than 10^{20} at/cm³. An enhanced oxidation rate at high doping levels has already been observed in the case of classical oxidation on c-Si substrates. It is explained as a Si-vacancy contribution mechanism in the silicon substrate which may provide reaction sites for a chemical reaction converting Si to SiO_2 and consequently increase the rate at which this reaction occurs [17]. In addition, it seems that as for oxide kinetics on standard virgin Cz-grown Si substrate, the polyoxide thick-

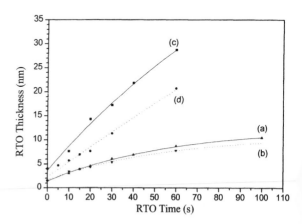

Fig. 1. RTO growth kinetics of oxides grown on standard virgin c-Si and poly-Si covered substrates at oxidation temperatures of 1000°C: (a) RTLPCVD in-situ PH_3-doped poly-Si films ([P] = 1.5×10^{20} at/cm³), (b) p-type (100) oriented Cz-grown Si (8–15Ω · cm) and 1100°C: (c) LPCVD poly-Si films $POCl_3$-doped ([P] = 10^{21} at/cm³) and (d) same samples as (b).

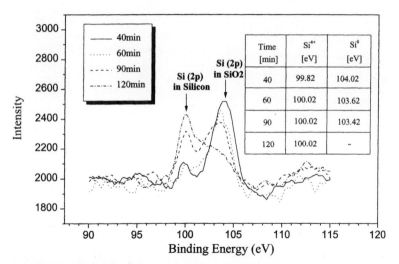

Fig. 2. Si-2p XPS spectrum of RTO polyoxide grown on in-situ PH₃-doped RTLPCVD polysilicon film. insert: binding energy shift as a function of the Ar⁺ sputtering time of an ultra-thin (6 nm) RTO polyoxide ($T = 1000°C/30$ s).

ness saturates at a process temperature of 1000°C and RTO time from about 60 s.

3.2. Chemical polyoxide structure

XPS analysis has been undertaken on thin RTO polyoxides layers in order to check their chemical structure. MgKα radiation was used as the excitation source and Ar⁺ ion was used for ion depth sputtering. An example of a 6 nm thick polyoxide XPS spectrum is given in Fig. 2. Si^{4+} (100 eV) and Si^0 (104 eV) peaks correspond usually to the Si-2p binding-energy level of Si atoms in Si substrate and in the silicon diox-

ide, respectively. This study has revealed intermediate binding states (Si^+, Si^{2+}, Si^{3+}) corresponding to a Si-rich polyoxide (SiO_x, $1 < x < 2$) structure which is extended in the vicinity of the poly-Si/polyoxide interface. In addition, as shown in the insert, the Si^{4+} binding-energy shift with pulverisation time proves some structure defects at the polyoxide/poly-Si interface which are likely due to dangling bonds and distorted O–Si–O bonds. This characteristic feature has already been pointed out for RTO oxide grown on c-Si substrates [16]. Furthermore, it can be noted that, within the instrumental accuracy, no visible P-2p signal was detectable for thin polyoxide layers.

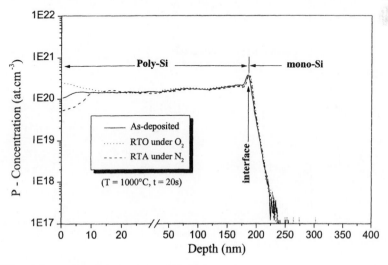

Fig. 3. SIMS phosphorus profiles of in-situ PH₃-doped RTLPCVD poly-Si films: (a) as-deposited, (b) after RTO at 1000°C 30 s and (c) after RTA under N₂ at 1000°C/30 s.

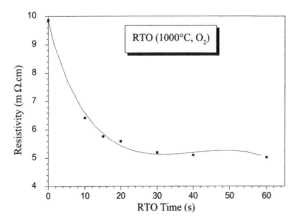

Fig. 4. In situ PH$_3$-doped RTLPCVD poly-Si (200 nm) sheet resistivity versus RTO time treatment.

3.3. Phosphorus SIMS depth profile

In Fig. 3, SIMS phosphorus profiles are given for in situ PH$_3$-doped RTLPCVD poly-Si films: (a) as-deposited, (b) after RTO at 1000°C/20 s and (c) after RTA under N$_2$ at 1000°C/20 s. It can be observed that the polyoxide cap blocks dopant outdiffusion from the poly-Si surface. Moreover, four-point probe resistance measurements indicated that the sheet resistivity is around 2.5 times as high when RTA is implemented under N$_2$ instead of O$_2$ atmosphere owing to surface dopant loss. In addition, a surface phosphorus pile-up at the poly-Si/polyoxide interfaces is noted. According to some authors, local pileup of dopants at grain boundaries leads to excessive incorporation of dopants in an oxide [4, 18]. Thereby, the polyoxide layer would

present poor dielectric properties identical to P-glass oxide. However, as mentioned above, the P-2p peak was undetectable by XPS analysis.

3.4. Electrical measurements

We can see from Fig. 4 that the RT-LPCVD polysilicon emitter sheet resistance decreases when the oxidation duration increases up to 30 s and saturates from this value owing primarily to dopant activation. Grain size growth is limited by the poly-Si emitter thickness and phosphorus dopant concentration. Indeed, Wada and Nishimatsu [19] have shown that only phosphorus doping in excess of 4×10^{20} at/cm^3 is capable of enhancing grain growth under high temperature annealing. In addition, Fig. 5 displays the high frequency C–V curve for a moderately (2×10^{17} at/cm^3) POCl$_3$-doped poly-Si film deposited on a 200 nm thermally oxidized Si substrate. This doping level was chosen in order to observe an ordinary C–V curve with depletion regime and extract some information concerning electrical polyoxide quality. To eliminate backside-contact parasitic resistance and capacitance, a Hg-probe contact configuration was employed. The flatband voltage (V_{fb}) deduced from the C–V curve analysis gives a total positive charge density in the polyoxide layer of approximately 10^{11} cm^{-2}. This value is similar to those found in RTO oxides grown on c-Si substrates [16]. A relatively low dielectric breakdown field of around 4 MV/cm is measured from I–V characteristics considering that breakdown voltage occurs when the current density reaches 1 µA/cm^2. An identical value was obtained for a phosphorus concen-

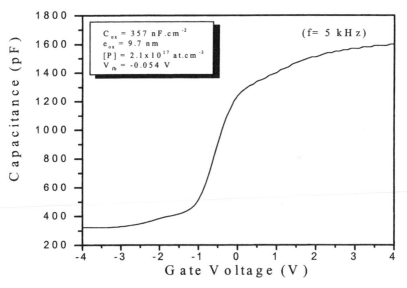

Fig. 5. C–V characteristics (5 kHz) for RTO polyoxide (10 nm) grown on moderately POCl$_3$-doped polysilicon layer ([P] = 2×10^{17} at/cm^3).

tration below 4×10^{20} at/cm^3 [4]. The relatively low dielectric strength of polyoxide is generally attributed to local high electric field associated with roughness and asperities at the poly-Si surface [8].

4. Summary

We have investigated RTO polyoxide growth and particularly effects of oxidation temperature and high phosphorus doping level on oxidation kinetics. Oxidations were carried out both on classical POCl$_3$-doped LPCVD polysilicon films and in situ PH$_3$-doped RTLPCVD poly-Si layers in order to point out the integrated multiprocessing possibility offered by RTP treatments.

It can be noted that growth kinetics are more accelerated as the oxidation temperature increases and even more when the phosphorus dopant concentration is greater than 10^{20} at/cm^3.

However, sheet resistivity measurements and SIMS analysis showed that the RTO polyoxide can play an effective role as both a cap layer blocking phosphorus outdiffusion and a surface passivation layer. This process scheme may permit tight control of the poly/polyoxide interface and thereby insure a good device performance and reproducibility.

Acknowledgements

Authors express their thanks to Ch. Dubois for SIMS analysis and J. Lebrusq for XPS measurements. This work received a financial support by the European Commission under the contract No. JOUR3 CT 95 0069.

References

[1] Sunami H. J Electrochem Soc 1978;125(6):892.
[2] Saraswat KC, Singh H. J Electrochem Soc 1982;129(10): 2321.
[3] Lemiti M, Audisio S, Mai C, Balland B. Rev Phys Appl 1989;24:133.
[4] Shih W, Wang C, Chiao S, Chen L, Wu N, Batra T. J Vac Sci Technol A 1985;3(3):967.
[5] Irene EA, Tierney E, Dong DW. J Electrochem Soc 1980;127(3):705.
[6] Carim A, Sinclair R. J Electrochem Soc 1987;134:741.
[7] Eernisse EP. Appl Phys Lett 1979;35:245.
[8] LeBlanc DG, Tarr NG, Ellul JP, Tay SP, Theriault RE. J Electrochem Soc 1992;139(1):227.
[9] Ohyu K, Wade Y, Iijima S, Natsuaki N. J Electrochem Soc 1990;137(7):2261.
[10] Nulman J. Mater Res Soc Symp Proc 1987;74:641.
[11] Zhang KX, Osburn CM, Hames G, Parker C, Bayoumi A. J Electrochem Soc 1996;143(2):744.
[12] Slaoui S, Hartiti B, Busch MC, Muller JC, Siffert P. Mater Res Soc Symp Proc 1991;387:224.
[13] Sivoththaman S, Laureys W, Nijs J, Mertens R. Mater Res Soc Symp Proc 1995;387:259.
[14] Kallel S et al. Paper presented at this conference.
[15] Gonon N, Gagnaire A, barbier D, Glachant A. J Appl Phys 1994;76(9):5242.
[16] Semmache B, Lemiti M, Gagnaire A, Laugier A. 14th European Solar Energy Conference and Exhibition. 1997. p. 1404.
[17] Ho CP, Plummer JD. J Electrochem Soc 1979;126(9): 1516.
[18] C Chang C, Sheng TT, Shankoff TA. J Electrochem Soc 1983;130(5):1168.
[19] Wada Y, Nishimatsu S. J Electrochem Soc 1978;125(9): 1499.

PERGAMON

Materials Science in Semiconductor Processing 1 (1998) 281–285

MATERIALS
SCIENCE IN
SEMICONDUCTOR
PROCESSING

Rapid thermal oxidation of porous silicon for surface passivation

L. Debarge [a],*, J.P. Stoquert [a], A Slaoui [a], L. Stalmans [b], J. Poortmans [b]

[a]CNRS-PHASE (UPR 292), 23 rue du Loess, 67037 Strasbourg, France
[b]IMEC vzw, Kapeldreef 75, 3001 Leuven, Belgium

Abstract

Rapid thermal oxidation with dry oxygen has been carried out on porous silicon (PS) films formed by electrochemical etching. The purpose of the paper was to investigate the surface passivation capability of the oxidized PS layers and to understand the oxidation mechanism. Rutherford back scattering (RBS) and X-ray photoemission spectroscopy (XPS) analyses confirmed the formation of a stoichiometric quasi-silicon dioxide. Besides, elastic recoil diffusion analysis (ERDA) demonstrated that a high concentration of hydrogen is still present in the PS film even after oxidation. RTO resulted in a good surface passivation effect at high temperature (> 1000°C) as seen by internal quantum efficiency analysis. However, lifetime in bulk silicon is affected by the RTO process. © 1999 Elsevier Science Ltd. All rights reserved.

1. Introduction

Porous silicon is used for its photo- and electro-luminescence properties [1], as anti-reflection coating (ARC) [2,3], as a gettering layer for low-cost epitaxial substrates [4], in photodetectors [13] and in many others electronic devices [5]. However, the large amount of highly reactive internal surfaces of porous silicon (PS) compared to Si increases drastically its sensitivity to ambient air. Thus, the use of as-grown porous silicon in electronic devices is limited because material degradation occurs with time. To overcome this problem, a post-(electro)chemical etching treatment is needed to passivate and/or oxidize the porous silicon to avoid time-dependent properties. Several ways have been presented in recent literature to oxidize porous silicon: with wet and dry O_2 flux [6], high and low temperature processes, O_3, H_2O_2 [7], nitridation [8], etc. Although PS showed fragility to high-temperature processes, the advantages of rapid thermal oxidation treatment have already been put forward [9], either for luminescence effects or passivation utilization [10]. Indeed, a short exposure to light (less than 1 min) can preserve the nano- (or micro-) crystalline structure of PS [9].

In this work, we investigate the effects of rapid thermal oxidation (RTO) on a thin yellow PS layer formed on a highly doped n^+ emitter. Our purpose is to apply porous silicon as an anti-reflection layer for solar cells. In addition, the RTO process should passivate the porous silicon, and also conserve its anti-reflection and light diffusion properties. We first focus on the material properties, to understand the rapid oxidation mechanism. The distribution of oxygen and hydrogen in the RTO-PS are evaluated, respectively, with RBS (Rutherford back scattering), XPS (X-ray photoemission spectroscopy) and ERDA (elastic recoil diffusion analysis) measurements. Then, the evolution of the optical properties is studied by reflectance measurements. The previous analyses lead to a better understanding of the RTO passivation possibilities, visualized by internal quantum efficiency (IQE) at 400 nm. Furthermore, we discuss the effect of RTO on minority carriers lifetime in the Si-bulk.

* Corresponding author. Tel.: + 33-3-8810-6337; fax: + 33-3-8810-6335; e-mail: debarge@phase.c-strasbourg.fr

2. Experiment

The base material is a 600 μm thick wafer, (100) oriented B-doped Cz-Si on which an n^+–p junction is produced by diffusion from P_2O_5 solid sources in an open-tube furnace. The PS samples are prepared by conventional electrochemical anodization and the backside of the cell is covered with a 2 μm evaporated Al layer serving as ohmic contact. Mirror-polished (MP) wafers having the same emitter characteristics were also used as a reference. Before the RTO process, MP and PS samples are dipped in a 2% HF solution for 30 s in the dark.

Our RTP unit is a commercial stainless-steel furnace JIPELEC (Grenoble, France) heated from the top by twelve halogen lamps and being water-cooled. The RTO process is performed under a dry oxygen flux (10 l/min) at temperatures between 700 and 1100°C for 30 and 60 s. The ramping rates are 150°C/s when heating and 80°C/s on cooling. The samples were held on quartz pins and the temperature control was made with the lamps power control by means of an appropriate calibration. The presence of Al on the backside makes a close-loop control based on the use of a pyrometer impossible.

The calibration, using attached thermocouples, has been performed to correlate the lamp power with the wafer temperature for Si(MP), Si(MP)/Al and PS/Si/Al samples. A temperature difference of about $100 \pm 10°C$ is observed between a MP-Si and a PS/Si/Al sample over the whole range of annealing temperature. This temperature enhancement can be explain by (i) the low emissivity of the Al–Si eutectic at the back of the cell (formed at 577°C) which reduces the radiation energy loss [11] and (ii) the low reflectivity of PS on the top of the wafer compared to MP-Si which induces more light penetration. This calibration allowed us to work with an equivalent wafer temperature for MP reference and PS samples.

Different techniques were used for PS analysis before and after oxidation. The hydrogen and oxygen content in PS were, respectively, determined by ERDA, RBS and XPS measurements. For the ERDA, a 2.9 MeV α-particles beam was employed and RBS measurements were accomplished using a 2 MeV α-particle beam, with a 150° retrodiffusion angle. Reflectance measurements were performed with a Lambda 19 (Perkin Elmer) spectrometer using a 60 mm diameter integration sphere. Lifetime measurements were obtained by photoconductivity decay (PCD) analyses, after removal of the emitter and the back surface field created by the aluminium diffusion during RTO.

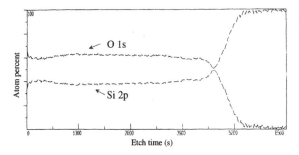

Fig. 1. XPS spectrum of PS after 800°C RTO.

3. Results and discussion

The knowledge of the PS morphological characteristics is important to understand the rapid thermal oxidation mechanism. It has been shown from XTEM pictures in previous work [10], that the used porous silicon presents a porosity of about 70% and is formed by a tangle of Si-wires of 8 to 10 nm diameter.

3.1. Oxygen and hydrogen distribution

Oxygen incorporation has been studied by RBS and XPS. Fig. 1 is an XPS depth profile showing the Si2p and O1s photoemission lines in the PS layer after a RTO treatment at 800°C. The PS/Si interface is reached after 5000 s etch time. The XPS results show a constant O/Si ratio of 1.5 throughout the whole RTO-PS layer for process temperatures above 800°C. This ratio of 1.5 is due to preferential sputtering of oxygen in XPS and corresponds to a stoichiometric SiO_2. This constant ratio demonstrates an homogeneous incorporation of the oxygen in the layer. However, at 700°C, the O/Si XPS ratio reaches only 1.4, corresponding to an incomplete oxidation of the PS, as well confirmed by RBS (Fig. 2). Material parameters such as layer thickness (t_{PS}) and O/Si ratio can be integrated in an RBS-evaluation program to fit the experimental results. Fig. 2 shows one RBS spectrum of a sample treated at 800°C for 30 s and the related calculated data curve. The spectrum of as-grown PS is also included for comparison. t_{PS} is the thickness of PS considering a void fraction in PS equal to zero. The height of the O-related peak is proportional to the ratio O/Si. Analyses of the as-grown PS indicates the presence of Si suboxides due to its high reactivity to ambient air. For RTO temperatures higher than 800°C, the related RBS calculations confirmed a O/Si ratio equal to 2, proving the transformation of PS in a quasi-SiO_2. The PS/Si interface position on the RBS spectra shifts towards lower channels number. This can be correlated with the t_{ps} increase (calculation results in Fig. 3) and demonstrates a densification of the PS layer and the oxidation of the Si substrate

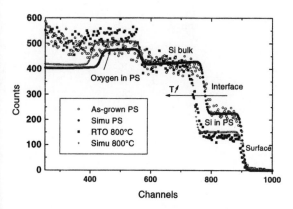

Fig. 2. RBS spectrum and calculated data for RTO 800°C, 30 s and as-grown PS.

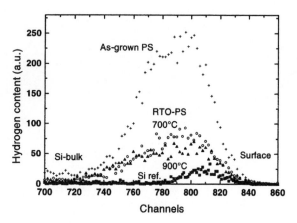

Fig. 4. Hydrogen content of Si, PS and RTO-PS.

underneath as witnessed by XTEM observations as well [10]. An increase of t_{PS} up to 20% for increasing RTO temperatures is demonstrated. Direct oxidation of highly doped mirror polished silicon are resulting in oxide layers thickness of 10 to 20 nm for comparable RTO conditions [12]. These thickness values are very close to those deduced from RBS simulations and confirm the RBS results.

A high temperature treatment induces Si–Si bond breaking, occurring after Si–H bond breaking and oxygen adsorption by the dangling bonds, resulting in PS volume enhancement due to a structural rearrangement [5]. At 700°C, oxygen adsorption occurs easily on the Si-wires due to their high surface energy. We can assume that PS keeps its wire structure after a low temperature RTO treatment. The resulting silicon wires are then thinner with a larger energy bandgap [13]. XPS spectra already show a complete oxidation at 800°C. At higher temperatures than 800°C, the PS undergoes a complete oxidation in a few seconds and the oxidation process continues on the bulk-Si with oxygen diffusion through SiO_2, creating an intermediate oxide layer. The oxidation then becomes diffusion-limited. The high surface energy of porous silicon, favoring oxygen adsorption, strongly enhances the oxidation kinetics and reduces the duration of the reaction-limited step when applying high temperature treatments.

The hydrogen distribution in the PS layer before and after RTO detected by ERDA technique is shown

RTO conditions (30s)	t_{PS} (nm)
No RTO	100
700°C	110
750°C	105
800°C	115
900°C	120

Fig. 3. Effective PS thickness t_{PS} versus RTO conditions.

Fig. 4. The as-grown sample presents a high concentration of hydrogen due to the large surface area of PS containing a large amount of SiH_x bonds. ERDA measurements still show a high hydrogen concentration in RTO-PS. About one third to one fourth of the hydrogen existing in as-grown PS is still present after RTO. This can be explained by the hydrophilic character of oxidized silicon and by the high amount of internal surfaces in oxidized PS encountered by the α-particles beam. Previous FTIR analyses [14] also proved that some specific Si–H bonds need a high temperature process (> 800°C) to be broken. Calculations based on ERDA results show that the H concentration is not constant over the PS layer after oxidation. The hydrogen peak can be deconvoluted and we find that the hydrogen content reaches 30% at the surface and decreases to zero towards the PS/Si interface. The hydrogen content also provides information on the RTO-PS morphology since the hydrogen can be correlated to still existing pores in RTO-PS because of its hydrophilic character. Even though the oxidation of the porous layer is complete, the original porosity (70%) causes the final structure to be a porous oxide with a decreasing porosity towards the PS/Si interface.

3.2. Reflectance measurements

The reflectance spectra of PS-samples before and after oxidation at different temperatures are shown in Fig. 5. The PS reflectance spectrum presents a maximum peak at 560 nm (yellow) and two minima at 430 and 1010 nm, corresponding to destructive interference. The reflectance increases beyond 1050 nm, due to the presence of Al on the backside of the cell. The RTO treatment changes the optical properties of the initial abrupt Si–Al interface, forming an Si–Al alloy which reflects less light.

The reflectance characteristics shift from a two-minima into a single minimum one. The maximum PS

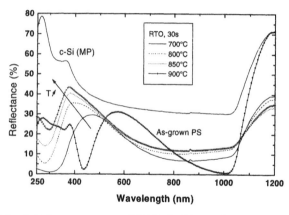

Fig. 5. Reflectance spectra of RTO-PS (700–800°C) compared to as-grown PS.

RTO condition (30s)	IQE MP (%)	IQE PS (%)
No RTO	30	19
700°C	29	20
800°C	34	28
850°C	39	28
900°C	34	31
1000°C (60 s)	/	45
1100°C	/	60

Fig. 6. Spectral response at 400 nm for mirror-polished (M and porous Si (PS) surfaces.

reflectance peak at 560 nm also shifts towards shorter wavelengths with increasing temperature. These effects on reflectivity can be explained by the fact that SiO_2 is a less dense optical medium, which lowers its refractive index. As it can be seen from ellipsometric spectroscopy [10], the refractive index shifts from ~2 to ~1.5 after RTO, which is unfavorable for ARC applications. Furthermore, with increasing RTO temperature, the peak corresponding to the first direct transition of silicon appears (at 365 nm), showing the bulk Si underneath and demonstrates that the RTO-PS becomes transparent to UV-light. The extinction coefficient k becomes zero, showing no absorption anymore in the RTO-PS for process temperatures above 800°C, confirming its complete oxidation.

3.3. Surface passivation

To test if the RTO process passivates porous silicon surface states, internal quantum efficiency (IQE) data of solar cells with and without PS were compared in the short wavelength region before and after oxidation. IQE was deduced from the external quantum efficiency (EQE) divided by $(1 - R)$ where R is the reflectivity. IQE results at 400 nm are reported in Fig. 6, comparing PS to mirror polished samples. The IQE values at 400 nm increase continuously with temperature both for PS and MP samples. The values for PS samples are systematically lower than those of MP samples, showing the difficulty in passivating porous silicon surfaces. However, a good surface passivation is reached for a process temperature above 1000°C only.

The IQE results for as-grown PS and the 700°C RTO sample are comparable. At 700°C, as already shown in Section 3.1, the oxidation is incomplete and a large amount of surface states acting as recombination centers are still present in PS. The high amount of dangling bonds creates absorption losses in the porous

silicon layer. After RTO treatment above 800°C, su face states will be saturated by oxygen after the hydr gen desorption. As the temperature process increase the original wires forming PS become thinner, resulti in a porous silicon layer with a larger bandgap enerς The increased blue response at a temperature abo 1000°C is attributed to an oxidation of the bulk si con, passivating perfectly the PS/Si interface.

3.4. Bulk lifetime measurements

A degradation in the red response was observed the IQE results in addition to the enhancement in t blue response. This can be an indication of bulk silic degradation and/or a loss of the diffusing character porous silicon. Minority carriers lifetime measuremeι were performed on the samples using the photocondι tivity decay technique (PCD). The samples were dipp in a HF solution to reduce surface recombinatio Results reported in Fig. 7 compare the effect of RΊ on lifetime values on PS and MP samples. A stro degradation of the effective lifetime is observed even a low temperature (700°C) and slightly decreases higher temperatures.

Lifetime degradation is usually attributed to pui tual (often metallic impurities) or extended defects (c locations) in the material. The observed lifeti decrease cannot be attributed to diffusion of impurit imported from porous silicon, as no significant diff ence is observed between samples with a PS layer an MP-surface. The observed degradation can then attributed to the activation of internal metallic impι ties due to the fast quenching step in RTO [15].

RTO conditions (30s)	Lifetime (µs) PS	Lifetime (µs) MP
no RTO	47	47
700°C	19.3	26.4
800°C	16.6	12.4
850°C	24.7	13.2
900°C	13.9	15.2

Fig. 7. Evolution of Si lifetime with RTO treatments.

4. Conclusion

A calibration was accomplished in order to apply temperature equivalent RTO treatment to samples with a PS or an MP surface. RBS, XPS and ERDA measurements demonstrate that hydrogen is still present even after an RTO at elevated process temperatures and that the oxidation is not complete at RTO temperatures below 800°C. Oxidation results in a volume enhancement due to structural rearrangement after hydrogen desorption. RTO-PS presents good passivation properties for a process temperature above 1000°C because of the formation of an intermediate oxide layer passivating the PS/Si interface. However, high temperature processes induce the activation of metallic impurities resulting in a lifetime degradation of the bulk silicon.

Acknowledgements

The authors are strongly indebted to Dr. T. Conard for performing XPS measurements. This work is funded by the EU in the frame of the Joule III BANSIS-project under contract number JOR3-CT960109 and by IWT (Flemish Institute for Promotion of Scientific-Technological Research in the Industry).

References

[1] Fauchet PM. J Lumin 1996;70:297.

[2] Bilyalov RR, Lautenschlager H, Schetter C, Schomann F, Schubert U, Schindler R. 14th European Photovoltaic Solar Energy Conference and Exhibition. 1997. p. 788.

[3] Strehlke S, Sarti D, Krotkus A, Grigorias K, Lévy-Clément C. Thin Solid Film 1997;297:291.

[4] Tsuo YS, Menna P, Pitts JR, Jantzen KR, Asher SE, AL-Jassim MM, Ciszek TF. 25th IEEE Photovoltaic Specialists Conference. 1996. p. 461.

[5] Arita Y, Kuranari K. Jpn Appl Phys 1997;36 (Part 1)(3A):1040.

[6] Chen H, Hou X, Li G, Zhang F, Yu M, Wang X. J Appl Phys 1996;79(6):3282.

[7] Frotscher U, Rossow U, Ebert M, pietryga C, Richter W, Berger MG, Arens-Fischer R, Münder H. Thin Solid Film 1996;276:36.

[8] Li G, Hou X, Yuan S, Chen H, Zhang F, Fan H. J Appl Phys 1996;80(10):5967.

[9] Petrova-Koch V, Muschik T, Kux A, Meyer BK, Koch F, Lehmann V. Appl Phys Lett 1992;61(8):943.

[10] Stalmans L, Poortmans J, Bender H, Jin S, Conard T, Debarge L, Slaoui A, Nijs J. 1st Porous Semiconductors-Science and Technology Conference, Meeting Abstracts. Mallorca, Spain, 1998. p. 30–1.

[11] Ventura L, Noël S, Lachiq A, Slaoui A, Muller JC. Proc. 14th EPSECE. 1997. p. 108.

[12] Slaoui A, Hartiti B, Busch MC, Muller JC, Siffert P. MRS Conf. Proceedings Symp. Rapid Thermal Processing and Integrated Circuits. 1990. p. 224 and 409.

[13] Lee MK, Wang YH, Chu CH. IEEE J Quant Elec 1997;33(12):2199.

[14] Schaefer A, Frankel D, Stucki F, Gopel W, Lapeyre GJ. Surf Sci 1984;139:L209.

[15] Eichhammer W, Vu-Thong-Quat, Siffert P. J Appl Phys 1989;66(8):3857.

PERGAMON

Materials Science in Semiconductor Processing 1 (1998) 287–292

MATERIALS
SCIENCE IN
SEMICONDUCTOR
PROCESSING

Deposition and crystallization of a-Si thin films by rapid thermal processing

S. Girginoudi *, D. Girginoudi, N. Georgoulas, A. Thanailakis

Department of Electrical and Computer Engineering, Democritus University of Thrace, 67100 Xanthi, Greece

Abstract

The deposition and crystallization of a-Si thin films grown by rapid thermal processing have been studied, using transmission electron microscopy. The a-Si films were deposited in a rapid thermal processor at reduced pressures in the temperature range of 530–580°C, at different deposition pressures and silane flow rates and subsequently were annealed in-situ by high temperature rapid thermal annealing (RTA) or by a two-step annealing process involving low temperature furnace annealing (FA) followed by high temperature RTA. The activation energy of a-Si deposition was found to be approximately 1.7 eV, in reasonable agreement with the conventional LPCVD technique. It has been found that the deposition temperature and deposition rate have a strong effect on the grain size, which is attributed to the nucleation processes in the bulk of the films. The combination of low deposition temperature, high deposition rate and a two-step annealing process permits the low temperature growth of poly-Si films of 100 nm thickness, with large grains of 520 nm size, containing a low density of microtwins and characterized by very low surface roughness of 2.2 nm. © 1999 Elsevier Science Ltd. All rights reserved.

1. Introduction

The growth of poly-Si films on glass substrates is one of the most important processes in the fabrication of thin film transistors (TFTs) for active matrix liquid crystal displays [1–3]. For such applications, low temperature processes should be utilized to avoid glass substrate degradation, due to thermal damage, and the control of material quality properties, such as grain size, in-grain defect density and surface roughness is essential. The most widely used poly-Si growth technique is the solid-phase crystallization (SPC) of a-Si, deposited by low-pressure chemical vapor deposition (LP-CVD), in a conventional furnace, at temperatures not exceeding 600°C [2, 4–6]. However, this crystallization process has limitations, due to the long annealing time and the in-grain defect formation [7, 8].

On the other hand, the growing demand for smaller thermal budget, higher throughput and process automation and control led to the development of rapid thermal processing (RTP) techniques, where halogen lamps are used as a heating source. Moreover, RTCVD and RTP can be combined in-situ to deposit a-Si films at a low temperature and then anneal these at higher temperatures. This is accomplished by changing the ambient and pressure between lamp temperature cycles. Recently, RTA has been used as an alternative to the crystallization of a-Si films on glass substrates [9–11] and the results obtained in terms of device quality were similar to those obtained by conventional furnace annealing. However, only a few studies have so far been published in this area and also they cover a narrow range of experimental conditions.

In this work, we have investigated the deposition and crystallization of a-Si thin films grown by RT-LPCVD and subsequently annealed in-situ by RTA or by a two-step annealing process, using TEM. The deposition has been carried out in a cold-wall rapid thermal reactor for a wide range of deposition conditions. The crystallization of a-Si has been achieved using high temperature RTA at 850°C for 45 s, or a two-step thermal annealing process: the first step being a low temperature FA at 600°C for 6 h in order to grow

* Corresponding author.

1369-8001/99/$ - see front matter © 1999 Elsevier Science Ltd. All rights reserved.
PII: S1369-8001(98)00034-1

large grains containing high density of defects, mainly microtwins, and the second step annealing being a high temperature RTA at 850°C for 45 s, which is applicable to poly-Si films deposited even on soft glass substrates. This second annealing step was found to be very effective in eliminating the in-grain microtwins. Finally, the effect of deposition conditions and annealing parameters on the grain size and surface roughness will be reported and its nature will be discussed.

2. Experimental details

Amorphous Si films of 50–100 nm thickness were deposited on 4″ oxidized Si wafers by RT-CVD at reduced pressures, using the pyrolysis of SiH_4 diluted (20%) in Ar. Depositions were performed in a Jetstar of JIPELEC rapid thermal processor described elsewhere [8]. The system consists of a water cooled, stainless-steel, low-pressure (10^{-3} Torr) process chamber with quartz windows. The wafers were heated by 12 quartz halogen lamps. The temperature was monitored by an optical pyrometer focused at the center of the wafer. Prior to each set of experiments, the pyrometer was calibrated against a thermocouple attached to the wafers. The deposition of a-Si was carried out in the temperature range of 530–580°C and for a total pressure range of 0.93–3.8 Torr, obtained by changing the source gas total flow rate between 20 and 100 sccm. After the deposition, two sets of annealing experiments were carried out. In the first set, a-Si films were crystallized in-situ by RTA at 850°C for 45 s. In the second set, a two-step annealing involving a low temperature furnace annealing at 600°C for 6 h followed by RTA at 850°C for 45 s was applied.

Transmission electron microscopy was used to investigate the structural and morphological characteristics of the crystallized films. TEM samples were studied by a JEM120CX microscope.

3. Results and discussion

Fig. 1 shows the deposition rate of a-Si films as a function of the deposition temperature for different total pressures and source gas flow rates. From these Arrhenius plots it is obvious that the deposition process is surface-reaction limited, with activation energies of 1.75 and 1.67 eV for depositions performed at 2.8 and 2.3 Torr and source gas flow rates of 100 and 75 sccm, respectively. These results are in reasonable agreement with typical values obtained for polycrystalline Si deposition from silane in conventional LPCVD furnaces [5, 12].

The microstructure of as-deposited Si films is determined mainly by the deposition temperature, because

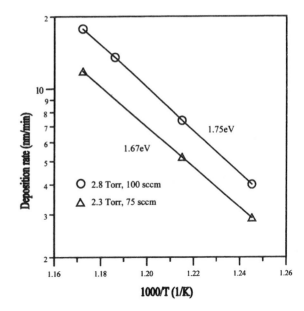

Fig. 1. a-Si film deposition rate versus the reciprocal deposition temperature for total pressures of 2.8 and 2.3 Torr and source gas flow rates of 100 and 75 sccm, respectively.

below a certain temperature the Si grows amorphous, whereas above this temperature it is polycrystalline. TEM studies showed that the Si films deposited at 630°C had small columnar poly-Si grains, which did not increase after thermal annealing and exhibited significant surface roughness. Si films deposited in the temperature range of 580–530°C were amorphous and exhibited a planar surface. Large grains with a high density of in-grain defects, namely microtwins having poorly defined grain boundaries, were formed by SPC after furnace annealing. A typical example of poly-Si films grown by SPC is shown in Fig. 2(a), which presents the TEM plan-view micrograph of a-Si films deposited at 530°C and crystallized by FA at 600°C for 6 h. The in-grain multiple twins cause a broadening of the diffraction rings, which is evident on the first ring in the related diffraction pattern shown in the inset of Fig. 2(a). In order to eliminate the in-grain defects, the FA samples were subjected to RTA for a very short time. This is possible because the activation energy of the twins is very low and they start to migrate at temperatures above 750°C. However, at this temperature the migration velocity of twin boundaries is very low. Therefore, higher RTA temperatures (e.g. 850°C) are strongly recommended in order to accelerate and annihilate the twin boundaries in a very short time, namely 45 s, which is applicable even to poly-Si films deposited on glass substrates. It should be pointed out that during this high temperature RTA no movement of grain boundaries is observed, because secondary recrystallization in poly-Si films starts at

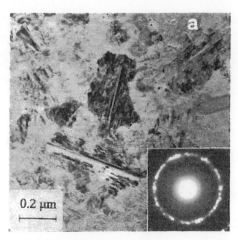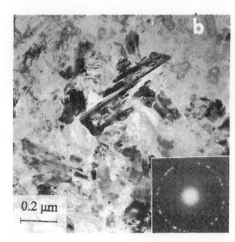

Fig. 2. TEM plan-view micrographs of: (a) 100 nm poly-Si film grown, from a-Si deposited at 530°C, by FA at 600°C for 6 h and (b) of the same film subjected to FA at 600°C for 6 h, followed by RTA at 850°C for 45 s.

temperatures above 1150°C. Fig. 2(b) shows the TEM plan-view micrograph of poly-Si films subjected to FA at 600°C/ 6 h and, subsequently, to RTA at 850°C for 45 s. The grain boundaries are sharp and the reduction of in-grain microtwins is also verified by the diffraction pattern, shown in the inset of Fig. 2(b), where a sharp first ring is evident, without forbidden spots owing to the double diffraction and diffraction lines, are observed in FA samples [8].

The effect of deposition conditions on the grain size, the density of the in-grain microtwins and the surface roughness of poly-Si films has been studied, in order to optimize the SPC poly-Si growth process using RTA at 850°C for 45 s or FA at 600°C/6 h followed by RTA at 850°C for 45 s. The structural characteristics of 100 ± 10 nm thickness poly-Si films, deposited at substrate temperatures (T_s) of 530, 550 and 580°C, with a constant deposition rate of 4.9 nm/min and annealed using the above two techniques, are shown in Figs. 3 and 4, respectively. It is obvious from the TEM

plan-view micrographs presented in Figs. 3 and 4 that, for both annealing techniques, the grain size clearly increases as the substrate temperature decreases from 580 to 530°C. The grains have an elliptical shape, with first order twins running mainly along the longest axis of the ellipse. The grain boundaries are sharp and the in-grain density of microtwins is very low, as it is obvious from the sharp diffraction rings of the diffraction patterns shown in the insets of Figs. 3 and 4.

For a given deposition temperature, the grains of samples subjected only to RTA are smaller in size than those of samples subjected to FA followed by RTA, as it is revealed by the plan-view micrographs shown in Figs. 3 and 4. This is attributed to the higher homogeneous nucleation rate obtained at higher annealing temperatures.

Fig. 5 shows the dependence of mean grain size on the deposition temperature of a-Si films of 100 ± 10 nm thickness, annealed by RTA at 850°C for 45 s or by FA at 600°C/6 h followed by RTA at 850°C for 45

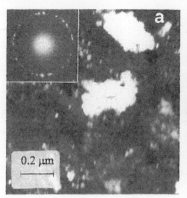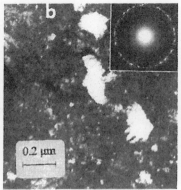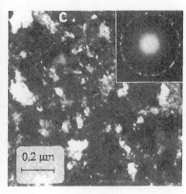

Fig. 3. TEM plan-view micrographs of Si films completely crystallized by RTA at 850°C for 45 s. The Si films were deposited at substrate temperatures: (a) 530°C, (b) 550°C and (c) 580°C, with the same deposition rate of 4.9 nm/min. The related diffraction patterns are shown in the insets.

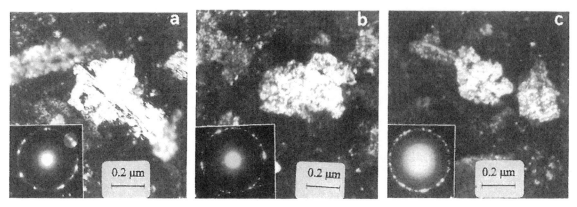

Fig. 4. TEM plan-view micrographs of Si films completely crystallized by FA at 600°C for 6 h followed by RTA at 850°C for 45 s. The Si films were deposited at substrate temperatures: (a) 530°C, (b) 550°C and (c) 580°C, with a constant deposition rate of 4.9 nm/min. The related diffraction patterns are shown in the insets.

s. The deposition rate was kept constant at 4.9 nm/min. The mean grain size of poly-Si increases as the deposition temperature decreases from 630 to 530°C, for both annealing techniques. Increasing the deposition rate by increasing the deposition pressure was also effective in increasing the grain size, as it is shown in Fig. 6. Since the crystallization temperature is kept constant, the crystallization growth rate is expected to be the same for all the samples deposited at different deposition conditions. Therefore, the different grain

sizes obtained by TEM studies on completely crystallized films are attributable to different nucleation rates during thermal annealing. TEM studies on Si films deposited at different substrate temperatures (T_s), with a constant deposition rate and annealed for different time intervals showed that there are more crystallites present at higher deposition temperatures. When the deposition temperature increases, the number of regions with increased order, which are favorable sites for nucleation during annealing increases, leading to

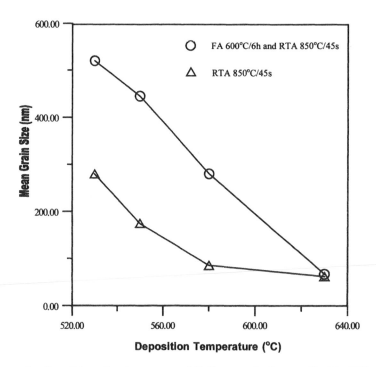

Fig. 5. Mean grain size as a function of deposition temperature of Si films completely crystallized by RTA at 850°C for 45 s or by FA at 600°C for 6 h followed by RTA at 850°C for 45 s. The Si films were deposited at a constant deposition rate of 4.9 nm/min.

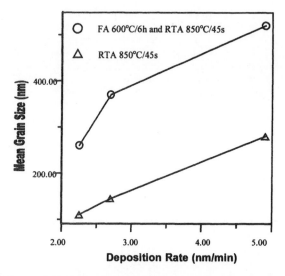

Fig. 6. Mean grain size of Si films deposited at 530°C and completely crystallized by RTA at 850°C for 45 s or by FA at 600°C for 6 h followed by RTA at 850°C for 45 s, as a function of the deposition rate.

higher nucleation rates, and thus to smaller grain size. At high deposition rates and low crystallization temperatures fewer nucleation sites are formed and thus larger grains can grow. Fig. 7 shows the TEM plan-view micrograph of 100 nm thickness films, deposited at 550°C and annealed at 600°C for 21 min. The film is partially crystallized, as can be seen from the coexistence of crystalline and amorphous phase, verified also by the electron diffraction pattern shown in the inset. The elliptical shape of the grains, which is related to the growth mechanism through multiple twins, is evident in the case when the crystallization of the film has

not been completed, where almost all the grains have elongated shapes.

The TEM measurements have shown that the surface roughness is very low (2.2–3.8 nm) and it does not change significantly with changes in the deposition conditions. Samples subjected to the two-step annealing process described above exhibit the minimum surface roughness.

4. Conclusions

In the present work, the effect of the deposition temperature, while maintaining the deposition rate constant, as well as the effect of the deposition rate, while maintaining the deposition temperature constant, on the grain size, the density of the in-grain microtwins and the surface roughness of poly-Si films have been studied using TEM, in order to optimize the SPC poly-Si growth process using RTA at 850°C for 45 s or FA at 600°C/6 h followed by RTA at 850°C for 45 s.

The two-step annealing process involving FA at 600°C/6 h followed by RTA at 850°C for 45 s, leads to large grains with low density of defects, mainly microtwins having abrupt grain boundaries, compared with the RTA process, which leads to small grains. The surface roughness of poly-Si films was found to be very low (2.2–3.8 nm). Samples subjected to the two-step annealing process exhibit the minimum surface roughness. In both annealing techniques, the mean grain size increases as the deposition temperature decreases, from 580 to 530°C, or when the deposition rate increases. This behaviour is attributed to the reduction of the nucleation rate during annealing as the deposition temperature decreases or the deposition rate increases. A mean grain size as large as 520 nm was obtained for films of a 100 nm thickness, deposited at 530°C, with a deposition rate of 4.9 nm/min and FA at 600°C/6 h followed by RTA at 850°C for 45 s.

Acknowledgements

The authors wish to thank Professor J. Stoemenos for his help in the TEM measurements and for stimulating discussions. The financial support from the Special account of Democritus University of Thrace (research program "Growth of polycrystalline silicon by Rapid Thermal Processing, RTP") is gratefully acknowledged.

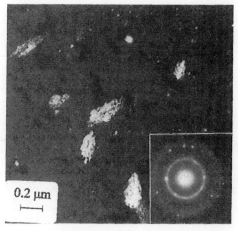

Fig. 7. TEM plan-view micrograph of Si film deposited at 550°C and annealed at 600°C for 21 min. The related diffraction pattern is shown in the inset.

References

[1] Noguchi T, Hiyashi H, Ohnishima T. J Electrochem Soc 1987;134:1771.

[2] Mimura A, Konishi N, Ono K, Owada JI, Hosokawa Y, Ono YA, Suzuki T, Miyata K, Kawakami H. IEEE Trans Electron Devices 1989;36:351.

[3] Wu IW, Chiang A, Fuse M, Övercoglu L, Huang TY. J Appl Phys 1989;65:4036.

[4] Kwizera P, Reif R. Appl Phys Lett 1982;41:379.

[5] Harbeke G, Krausbauer L, Stegmeier EF, Widmer AE, Kappert HF, Neugebauer G. J Electrochem Soc 1984;131:675.

[6] Hatalis MK, Greve DW. J Appl Phys 1988;63:2260.

[7] Haji L, Joupert P, Stoemenos J, Economou NA. J Appl Phys 1994;75:3944.

[8] Girginoudi D, Girginoudi S, Thanailakis A, Georgoulas N, Stoemenos J, Antonopoulos J. Thin Solid Films 1995;268:1.

[9] Stoemenos J, Economou NA, Haji L, Bonnel M, Duhamel N, Loisel B. In: Strunk HP, Werner JH, Fortin B, Bonnaud O, editors. Proceedings of the 3rd International Conference on Polycrystalline Semiconductors-Physics and Technology, vols. 37–38. Aedermannsdort, Germany: Scitec, 1994. p. 547.

[10] Plevert L, Haji L, Bonnel M, Duhamel N, Loisel B. In: Strunk HP, Werner JH, Fortin B, Bonnaud O, editors. Proceedings of the 3rd International Conference on Polycrystalline Semiconductors-Physics and Technology, vols. 37–38. Aedermannsdort, Germany: Scitec, 1994. p. 547.

[11] Bonel M, Duhamel N, Haji L, Loisel B, Stoemenos J. IEEE Electron Devices 1993;14:551.

[12] Foster A, Learn A, Kamins T. Solid State Technol 1986;29:227.

PERGAMON

Materials Science in Semiconductor Processing 1 (1998) 293–297

MATERIALS
SCIENCE IN
SEMICONDUCTOR
PROCESSING

The initial stages of Si thin deposits on foreign substrates in a rapid thermal chemical vapor phase reactor

D. Angermeier *, R. Monna, S. Bourdais, A. Slaoui

Laboratory PHASE (UPR 292 CNRS), BP 20, 67037 Strasbourg Cedex 2, France

Abstract

Nucleation and coalescence mechanisms of silicon clusters on silicon dioxide and high purity crystalline alumina substrates have been investigated in a rapid thermal chemical vapor deposition (RTCVD) system at atmospheric pressure for crystalline silicon thin film solar cells. Trichlorosilane (TCS) was employed as a reactant diluted in a hydrogen carrier gas. The nuclei density was determined as a function of the deposition time and the saturated nuclei density versus substrate temperature ranging from 900 to 1250°C for two different reactor configurations. Furthermore, the evolution of the average cluster size and variation and cluster density were studied to understand the impact of hydrodynamics and pre-annealing treatment of the substrate at deposition temperature before the nucleation step. We obtained the same induction periods for the establishment of steady state Si nucleation on both substrate materials. Conversely, the distinct physical properties of the two substrates influenced strongly the nucleation density, the nuclei and grains size evolution depending on the growth temperature and time. In this case, chemical reactions and surface conditions determine further ripening mechanisms and coalescence of the clusters resulting in coarser grains and faster layer growth for the Si deposits on alumina substrates. The effects of surface pre-treatments and operational parameters on the cluster and grain formation were analyzed by Nomarski and scanning electron microscopy (SEM). © 1999 Elsevier Science Ltd. All rights reserved.

1. Introduction

Rapid thermal chemical vapor deposition offers significant applications in microelectronic industry of in-situ multilayer-processing of different material composition (i.e. SiGe, III–V heterostructures, supperlattices) [1, 2]. The rapid thermal processing permits a quick temperature cycling to control the thermally driven surface reactions and generating a temperature field limited only to the wafer area which limits homogeneous reaction pathways in the reactor. Recently, the rapid thermal processing technique (RTCVD) at atmospheric pressure has become a potential candidate for generating thin film silicon layers in the range from 10 to 30 µm thickness within some minutes operating above 1000°C for thin film silicon solar cells application [3].

In this respect, the deposition of polysilicon material on foreign substrates as cost-effective carrier materials contributes significantly to the cost-reduction of thin film silicon solar cell and by guaranteeing long-term stability. Particularly, it is necessary to obtain large crystallites and a low volume of grain boundaries as recombination centers to limit the effect of minority carrier trapping and reduced effective diffusion lengths [4]. The objective is therefore to adapt the processing and control the grain size growth and evolution by the deposition conditions such as substrate temperature, chemical reaction and precursor input concentration. One of the means is to determine the decisive factors for the initial stage of growth, nucleation and its further progress up to the coalescence of the islands. According to Bloem, the deposit of silicon nuclei on amorphous SiO₂ or Si₃N₄ substrate would be

* Corresponding author. Tel.: + 33-3-8810-6337; fax: + 33-3-8810-6335; e-mail:angermeier@phase.c-strasbourg.fr

1369-8001/99/$ - see front matter © 1999 Elsevier Science Ltd. All rights reserved.
PII: S 1369-8001(98)00035-3

determined by the saturation nuclei density [5]. In any case, the dynamics of cluster evolution is complicated being subject to fundamental processes of surface diffusion, adsorption and desorption [6]. The adatoms arriving from the gas phase diffuse rapidly with low diffusion energy and encounter other atoms, producing nucleation of small stable clusters [7]. The small clusters will then merge into larger clusters of atoms which will increase further in size either via exchange of adatoms from the small clusters ('Ostwald ripening') or by simple coalescence [8].

In this report, we probed the rapid thermal CVD nucleation, stable cluster formation, coalescence and ripening process resulting in different grain sizes for Si on SiO_2 and on high purity crystalline Al_2O_3. We studied additionally the impact of the flow behavior in two different RTCVD reactor configurations on the nucleation process and its cluster size evolution and variation. The evaluation of the saturated nucleation density is combined with the nuclei size variation as a function of the temperature and deposition time. In addition, the growth rate and grain size evolution with the incubation time for Si layer formation on both substrates were analyzed.

2. Experimental

A schematic drawing of the employed RTCVD reactor configurations with top lamp heating are illustrated in Fig. 1. A bank of 12 tungsten–halogen lamps was used as a radiation source in this unit. The RTCVD system represents a horizontal single-wafer reactor consisting of a two piece chamber made of stainless steel and is rectangular in flow direction.

In the configuration 1, right above the silicon wafer the reactor chamber is equipped with an additional round-shaped elevation of 2 cm just beneath the quartz window while in configuration 2 the upper space alienated to the entire ceiling with no space left. The side walls and the top quartz window are water-cooled to elude heat memory effects and unwanted deposition. The deposition temperature is measured on the back side of the silicon wafer with a pyrometer. The silicon wafer (4 inches) serving as a susceptor is placed on four quartz pins to provide a better temperature uniformity and to minimize the thermal budget than the use of solid SiC-coated graphite susceptors.

In our experiments, the preparation of the alumina substrates prior to growth was implemented by an etching procedure in a diluted $HF:H_2$ (9:1) solution for 50 s and with a final DI water rinse whereas the SiO_2 samples were only water rinsed. Before loading the specimen into the reactor, the wafers were blown dry by N_2. Then the chamber was pumped down to a base pressure of 1 mTorr. Afterwards, H_2 was introduced to flush the chamber and establish a steady flow rate. In-situ cleaning was performed by a H_2 prebake at the growth temperature for 45 s before the Si deposition commenced at atmospheric pressure. For the Si deposit trichlorosilane ($SiHCl_3$) served as the source gas rarified in hydrogen carrier gas. The typical range of deposition temperature was set for the experiments from 900 to 1250°C.

In order to study the Si nucleation on SiO_2 and Al_2O_3, the variation of the operational parameters such as substrate temperature and deposition time were investigated. Materials analyses and deposition rate determination were performed by scanning electron microscopy (SEM) and Nomarski optical microscope.

3. Results and discussions

3.1. Si nucleation on SiO_2 substrate

The Si nucleation at high growth temperatures is critically influenced by the nature of the substrate and it is sensitive to the hydrodynamics of the gas flow and the rate of diffusion of the reactants. Fig. 2 gives an example of the silicon nuclei density on amorphous silicon dioxide at 1250°C with 15% TCS diluted in H_2. Moreover, the density of Si nuclei depends critically on the deposition time and on the RTCVD reactor configuration as shown in Fig. 3. In all cases, the density of nuclei reaches a saturation level and a subsequent decrease takes place due to the coalescence mechanism where the number of total stable clusters decreases with time. It can be unequivocally recognized that unfavorable flow pattern in reactor 1 [9] yields a lower

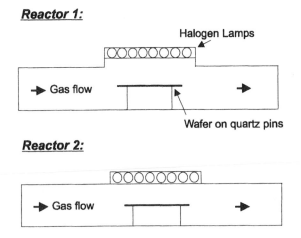

Fig. 1. Schematic diagram of the used RTCVD systems for nucleation and hydrodynamic studies and effects on the grain size evolution.

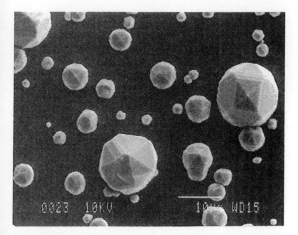

Fig. 2. Nomarski micrograph of Si nucleation on amorphous silicon dioxide at 1250°C.

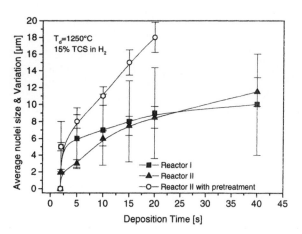

Fig. 4. The average nuclei size and variation versus the deposition time at 1250°C for two different reactor configurations.

nucleation density by almost one order of magnitude presenting a longer nucleation incubation period. Moreover, the start of the coalescence is retarded by the double time length than found in the compared reaction cell 2. This outcome underlines the fact that non-laminar flow patterns cause Si supersaturation in the gas phase and subsequent homogeneous nucleation depleting the partial pressure of the precursor. Secondly, it was observed that a temperature pretreatment at the growth temperature under H_2-flow in reactor 2 diminishes the nuclei density by one order of magnitude and the start of coalescence step. Whereas the average nuclei size evolution is crucially increased in comparison to the non-thermally treated specimen, as can be seen in Fig. 4. At this point, the larger nuclei sizes obtained are correlated to the decrease of the nuclei density per area. This might be due to the reaction of H_2 with the SiO_2 substrate surface promoting the adsorption of atomic hydrogen impeding the cover-

age of the substrate with the first silicon nuclei expressed by the following reaction [10]:

$$Si + O_{Substrate} + H_2(g) \longrightarrow SiO(g) + 2H_{ad} \quad (1)$$

However, the reduced density of nuclei are able to form a large cluster by directly capturing the adatoms from the gas phase and further at the expense of the just nucleated adatoms facilitating the overall Ostwald ripening mechanism [11].

In contrast to the difference in nuclei density in the two reactor concepts, the average nuclei size is less affected by the flow dynamics. However, instability in flow patterns by recirculations or non-laminar flow causes a broader distribution of cluster size and a subsequent local growth rate variation of the forming islands. Thus, nucleation behavior and layer formation is substantially determined not only by the operational parameters, but also by buoyancy driven flows or secondary flow that should be eluded in the reaction chamber [12].

Fig. 5 depicts the saturation density of the nuclei, N_S, dependent on the substrate temperature for the two reactor configurations. It can be noted that the saturated density increases with deposition temperature in a $SiHCl_3$–H_2 system, whereas the nuclei size of 6–9 µm depicts no change in temperature for both reactor configurations. However, this evolution of Si nuclei density stands in opposition to a SiH_4–HCl–H_2 system where the addition of $HCl(g)$ etches the substrate surface at higher deposition temperatures [13]. In our system, at below 1000°C there is a strong decrease of $SiCl_2$ concentration in the gas phase reducing the formation of Si atoms. Simultaneously, at lower substrate temperatures an increased adsorption of hydrogen atoms at the SiO_2 surfaces occurs hampering the nucleation of Si [5]. At elevated deposition temperatures the arriving adatoms overcome more easily the

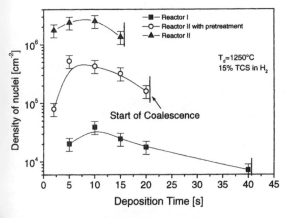

Fig. 3. Evolution of the nuclei density as a function of deposition time at 1250°C for two different reactor configurations.

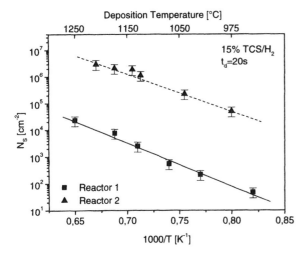

Fig. 5. Si saturation nuclei density on SiO_2 substrates after 20 s as a function of the substrate temperature for two different reactor configurations.

activation energy for surface adsorption and diffusion to generate stable clusters without the risk of evaporation into the gas phase [14].

3.2. Si nucleation on Al_2O_3 substrates

Next, the nucleation effect of Si deposit on high purity crystalline alumina substrates (3N) were carried out being subject to the growth temperature and duration. Fig. 6 represents the dependence of Si nuclei density versus the formation time. It can be observed that the density of clusters is larger for the Al_2O_3-substrate than that for the SiO_2 substrates in reactor configuration 1. Although the size of the nuclei in the two cases are quasi equal at 1200°C, the higher cluster density can lead to an earlier coalescence than for Si on

silicon dioxide. The long time for attaining the saturation density after about 15 s can be explained by the chemical reduction of the substrate by the silicon being a primary source of contamination shown by

$$2Si + Al_2O_3 \longrightarrow Al_2O(g) + 2SiO(g) \qquad (2)$$

This reaction may proceed under the conditions where the gaseous oxide of alumina may escape and prevent adsorption of Si adatoms on the surface [15].

On the other hand, the saturation density of Si nuclei on Al_2O_3 in the TCS–H_2 system decreases with increasing temperature as compared to the nucleation of Si on SiO_2 as shown in Fig. 7. Even throughout the temperature range 900–1250°C, the saturation Si nuclei density values on alumina are one order of magnitude above those of Si on SiO_2. Nonetheless, it should be pointed out that the reason for the reduction of the density at elevated temperatures is twofold. Hydrogen can suppress the reduction of TCS at the surface by forming the gaseous etchant HCl and simultaneously reacting with the substrate surface by generating $Al_2O(g)$ acting as an impurity source [16,17]. The appropriate chemical process can be written according to the following reactions:

$$SiHCl_3(g) + H_2 \longrightarrow Si + 3HCl(g) \qquad (3)$$

$$2H_2(g) + Al_2O_3 \longrightarrow Al_2O(g) + 2H_2O(g) \qquad (4)$$

Secondly, elevated mobilities of the small clusters and the adatoms merge exponentially in larger and stable clusters at high deposition temperatures above 1100°C for both reactor types. This can be recognized by the large cluster sizes of 10 µm at 1250°C and by subsequent decrease of the density of the smaller clus-

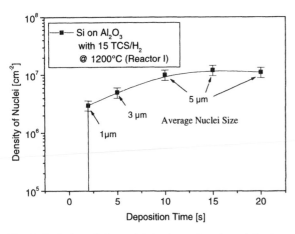

Fig. 6. Evolution of the nuclei density as function of the deposition time for Si on crystalline Al_2O_3 substrates at 1200°C.

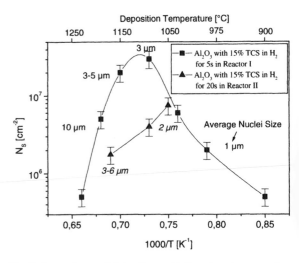

Fig. 7. Saturation nuclei density on Al_2O_3 substrate versus the substrate temperature for two types of reactor configurations with 15% TCS in H_2.

ters indicating an Ostwald ripening process for layer formation. Furthermore, the reinforced decrease in nuclei density in reactor 1 is supported by non-laminar flow patterns yielding supersaturation of Si and therefore gas phase nucleation and particle formation, but it does not influence the tendency of the ripening process.

3.3. Layer formation and grain size evolution of Si on Al_2O_3 and on SiO_2

In order to evaluate and compare the impact of the cluster size evolution on the layer growth and crystallite size progress for Si on Al_2O_3 and SiO_2 substrates, it is crucial to continue the Si deposition until large grains manifest. In Fig. 8, one can clearly identify that the greater nuclei density of Si on Al_2O_3 comprising the ripening process results in higher growth rates and larger grain sizes than for Si on SiO_2. In the latter case coalescence leads to a faster continuous layer formation, yet the reduced cluster density due to atomic hydrogen inhibition of the nucleation process prevents quick ripening by smaller clusters and impinging adatoms. Therefore, the ripening process seems to be an important growth mechanism in a RTCVD growth system where large nuclei can be constituted at temperatures above 1200°C merging into rapidly enlarging 3D islands which determine thereafter the grain size. These large entities are less mobile but grow faster into large grains at the expense of smaller ones and consequently increase the overall deposition rate.

4. Conclusion

We have conducted a thorough analysis of the Si nucleation process on crystalline alumina and silicon dioxide substrates in two distinct RTCVD reactor configurations at atmospheric pressure. The experiments have been carried out to determine the impact of the operational parameters and hydrodynamics on the Si nuclei formation, coalescence process and subsequent grain size evolution. It has been clearly demonstrated that the secondary flow patterns in one of the reactors is detrimental to the Si nucleation density for both types of substrates. For Si on alumina substrates higher cluster densities could be obtained than for Si on SiO_2 where the strong formation of atomic hydrogen with oxide on the substrate surface impeding an initial coverage by Si nuclei. Conclusively, the resulting smaller clusters have less chance of ripening to large islands merging into larger grains as it occurred, conversely, for the Si nucleation on alumina substrates.

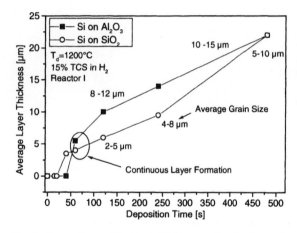

Fig. 8. Comparison of the layer thickness and grain size variation for Si on Al_2O_3 and SiO_2 substrates at 1200°C and with 15% TCS in H_2.

Acknowledgements

This paper was funded by the EU (JOULE-program) under contract No. JOR3-CT95-0080 and by ADEME-ECODEV, France.

References

[1] Green ML, Brasen D, Luftman H, Kannan VC. J Appl Phys 1989;65:2558.
[2] Vook DW, Gibbons JF. J Appl Phys 1990;67:2100.
[3] Faller FR, Schillinger N, Hurrle A, Schetter C. 14th EPVSEC. Barcelona, 1997. p. 784.
[4] Green MA. Silicon solar cells. In: Advanced principles and practice. Sydney: UNSW, 1995.
[5] Bloem J. J Cryst Growth 1980;50:581.
[6] Hirth JP. J Cryst Growth 1972;17:63.
[7] Venables JA. J Vac Sci Techol B 1986;4:870.
[8] Zinke-Allmang M, Feldman LC, Grabow MH. Surf Sci Rep 1992;16:377.
[9] Angermeier D, Monna R, Slaoui A, Muller JC. J Electrochem Soc 1997;144:3256.
[10] Claasen WAP, Bloem J. J Electrochem Soc 1980;127:194.
[11] Venables JA. Surf Sci 1994;299–300:798.
[12] Jensen KF. In: Hurle DTJ, editor. Handbook of crystal growth 3, Thin films and epitaxy, Part B. Elsevier Science Publisher, 1994. ch. 13.
[13] Claasen WAP, Bloem J. J Electrochem Soc 1980;127:1836.
[14] Stowell MJ. J Cryst Growth 1974;24–25:42.
[15] Cullen GW. J Cryst Growth 1971;9:107.
[16] Bloem J, Gilling LJ. In: Kladis E, editor. Current topics in materials science, vol. 1. North-Holland Publishing Co., 1978.
[17] Mercier J. J Electrochem Soc 1970;117:813.

PERGAMON

Materials Science in Semiconductor Processing 1 (1998) 299–302

MATERIALS
SCIENCE IN
SEMICONDUCTOR
PROCESSING

Growth and physical properties of in situ phosphorus-doped RTLPCVD polycrystalline silicon thin films

S. Kallel[a],[*], B. Semmache[a], M. Lemiti[a], Ch. Dubois[a], H. Jaffrezic[b], A. Laugier[a]

[a]*Laboratoire de Physique de la Matière, (CNRS, UMR-5511) Bât 502, Institut National des Sciences Appliquées de Lyon, 20 Avenue Albert Einstein, F69621 Villeurbanne Cedex, France*
[b]*Laboratoire MMP, Ecole Centrale de Lyon, F69131 Ecully Cedex, France*

Abstract

In situ phosphorus-doped (P-doped) polysilicon (poly-Si) thin films are obtained by rapid thermal low pressure chemical vapor deposition (RTLPCVD) in a single chamber RTP machine by using diluted silane ($SiH_4/Ar = 10\%$) and phosphine ($PH_3 = 200$ ppm). Deposition kinetics of poly-Si films were studied in the 600–850°C temperature range at fixed total pressure of 2 mbar and gas flow rate (100 sccm). Activation energy of 1.82 eV was calculated in the surface reaction deposition regime. Dopant activation has been obtained sequentially by RTO at 1000°C in pure O_2 atmosphere. This later process permits to both activate the phosphorus dopant and forms an ultrathin polyoxide which blocks dopant outdiffusion. Secondary ion-mass spectrometry (SIMS) analysis showed flat P-dopant profiles throughout the film thickness with a P concentration varying from 5.5×10^{20} to 2.4×10^{19} at/cm³ when the deposition temperature increases in the 600–850°C range. Grazing incidence X-ray diffraction (XRD) has been used to study the structural properties of the poly-Si layers. It appeared particularly that the amorphous to crystalline temperature transition occurs at around 650°C. Finally, four-point probe measurements showed that sheet resitivities in the mΩ cm range can be routinely achieved for in situ P-doped RTLPCVD poly-Si films. © 1999 Published by Elsevier Science Ltd. All rights reserved.

1. Introduction

LPCVD poly-Si thin films are largely used in integrated circuits technology for various applications. There are used as gate or interconnects in MOS devices and also as both a dopant source and a low resistance contact in high gain bipolar transistors fabrication [1–3]. Other applications include photovoltaic conversion [4, 5] and mechanical sensors [6]. For most, if not all, of these applications tight control of doping levels and the degree of the crystallinity of the layers is needed. Incorporation of dopants into poly-Si layers is usually made by means of a two-stage postdeposition process, which can be either a $POCl_3$ diffusion accompanied by an additional drive-in heat treatment or ion implantation, followed by activation annealing step at a temperature in-between 1000 and 1200°C [7]. In situ doping during LPCVD poly-Si deposition is an attractive technique insofar that it insures a flat doping depth profile of As-deposited layers and permits a process step economy. However, dopant activation is generally implemented ex situ in an appropriate reactor, whilst, RTLPCVD deposition with in situ doping of poly-Si films can be immediately followed by an RTA dopant activation treatment in the same reactor such that cross-contamination associated with wafer handling is avoided. In addition, low thermal budget requirements needed by the continuous scaling down in state of the art VLSI technologies could not be fulfilled by conventional furnace processing. Thereby, RTLPCVD deposition offers undoubtably a valid re-

---

* Corresponding author.

1369-8001/99/$ - see front matter © 1999 Published by Elsevier Science Ltd. All rights reserved.
PII: S 1369-8001(98)00036-5

sponse to develop a low thermal budget manufacturing scheme for several components.

In the present paper, RTLPCVD in situ P-doped poly-Si film deposition was studied. Structural and electrical properties were investigated as a function of the deposition temperature. P dopant incorporation mechanism during deposition is also discussed with respect to process and structural parameters.

2. Experimental

RTLPCVD poly-si layers were deposited on p-type (1–2 Ωcm) (100) oriented Cz-grown silicon 2 inches diameter wafers covered with 100 nm thick thermally grown oxide. Deposition were carried out in a FAV4 (Jipelec™) rapid thermal reactor which consists of a water-cooled, stainless-steel cylindrical chamber. Substrates are front-heated through a single quartz window by a bank of 12 tungsten halogen lamps. The temperature is closed-loop controlled by an optical pyrometer viewing the sample backside. A base pressure of about 10^{-4}–10^{-5} mbar can be attained within 10 min using a combination of mechanical and turbomolecular pumps. After usual degreasing and buffered-HF cleaning processes, the wafer is immediately loaded in the processing chamber. In situ P-doped RTLPCVD poly-Si thin films were prepared by pyrolysis of SiH_4/Ar (10%) mixture with adding 200 ppm phosphine gas at deposition temperature in the 600–850°C range and a total process pressure of 2 mbar. Total gas flow rate was kept at a fixed value of 100 sccm. In these conditions, the PH_3/SiH_4 mole ratio is fixed at 2.2×10^{-3}. A preheating step of the reactant gas at 350°C, just below the decomposition temperature of silane, was first applied in order to adjust the total process pressure then the ramp up step (100°C/s) starts toward the deposition temperature plateau at various hold time (30 s to 3 min).

3. Results and discussion

3.1. Polysilicon deposition kinetics

In Fig. 1 the deposition rate of RTLPCVD P-doped poly-Si vs. the deposition temperature is shown. Results of undoped series extracted from Ref. [8] are reported in order to put in evidence the phosphine species influence on the deposition rate. In the intrinsic series case, the Arrhenius plot identifies both a surface reaction-limited and a mass transport-limited regime already cited in classical LPCVD processes [9]. Transition between the two regimes occurs at 750°C and the activation energy for the surface-limited reaction is 1.7 eV (40 kcal mol). Besides, it appears that

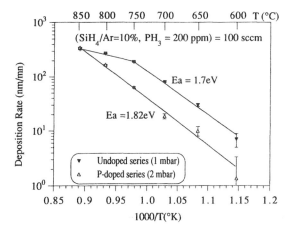

Fig. 1. Arrhenius plot of RTLPCVD in situ P-doped poly-Si deposition rates in the 650–850°C temperature range (P is 2 mbar). Effect of in situ P-doping.

the doped series displays only a surface reaction-limited regime with a slightly higher activation energy at around 1.82 eV (43 kcal/mol). This discrepancy is likely due to additional surface adsorption and desorption energies of SiH_4 source and H_2 byproduct species. Furthermore, it can be noted that despite a twice process pressure the deposition rate is notably reduced for the doped series. Indeed, as usually reported in the literature dedicated to classical LPCVD, phosphine species block the surface reaction sites and then inhibate the silane surface adsorption mechanism which control the poly-Si film deposition process [10, 11]. This feature appears to be less dominant at higher deposition temperatures when SiH_4 surface decomposition reactions are much more accelerated.

3.2. Structural analysis

Grazing XRD analysis were used in order to assess the degree of crystallinity, crystallite size and texture of RTLPCVD poly-Si films as a function of the deposition temperature. Only contribution of the major diffraction peaks, ie. (111), (220) and (311) was taken into account. Crystallite sizes, D_{hkl}, were calculated using the classical Sherrer formula [12]. Crystallographic orientation factors of poly-Si films were normalized with respect to an appropriate polycrystalline reference powder. XRD spectra showed that, as previously observed for the intrinsic series, the amorphous to crystalline temperature transition occurs at deposition temperature from about 650°C [13]. We have postulated that this discrepancy with respect to classical LPCVD films (580–600°C) is probably linked to the way by which the deposition is started. Gas-switching in classical furnaces allows previous residual water vapor thermal desorption of both substrates and internal wall, whereas temperature-swiching in RTP reactors does not. Moreover, it emerges that the structural amorphous phase contribution disappeared at deposition temperature from 750°C. Beyond this later temperature, the poly-Si films structure seems to be totally crystalline. Balanced (220) and (311) major crystalline orientations are also noted. Furthermore, as seen in Fig. 2, RTLPCVD poly-Si films presents a small-grain structure with an average grain-size of about 20 nm. Even though, grain-size enhancement is observed at deposition temperature from 800°C, when silane species surface mobility is less inhibited by the initial preferential phosphine adsorption mechanism. Note that after a subsequent RTA treatment (1100°C/ 20 s) under N_2 ambient, grain-size is slightly improved (30–40 nm). It seems that poly-Si grain size growth mainly depends on the initial structural state and is limited by the film thickness. Besides, according to the literature, high P-doping grain size growth mechanism could be effective only when dopant concentration is over 4×10^{20} at/cm^3 [14].

3.3. Phosphorus SIMS depth profiles

The phosphorus concentration vs. the deposition temperature is shown in Fig. 3 together with the P-dopant incorporation rate which corresponds herein to the polysilicon deposition rate and phosphorus concentration product. It appears that the phosphorus dopant concentration decreases from 5.5×10^{20} to 2.4×10^{19} at/cm^3 when the deposition temperature increases in the 600–850°C range. One may emphasize that the phosphorus concentration of 1.2×10^{20} at/cm^3 obtained at deposition temperature of 750°C corresponds approximately to the solid P solubility limit in

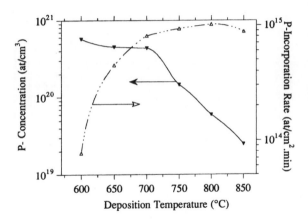

Fig. 3. SIMS phosphorus concentration together with incorporation rate vs. deposition temperature of RTLPCVD in situ P-doped poly-Si films.

Si at this temperature [15]. One may also remark that the PH$_3$/SiH$_4$ mole ratio is, by a mere chance, strictly respected at this deposition temperature. While, outside this deposition temperature, phosphorus concentration is either higher or lower than the solid solubility limit depending on the structural nature of the deposited layers. On one hand, the dopant concentration is relatively higher when the deposited films are amorphous owing to the disordered structure (dangling bonds). On the other hand, Si atom incorporation in the growing layer is more accelerated because the PH$_3$ decomposition rate is lower than one of SiH$_4$ at higher deposition temperature [11]. This behavior is well supported by the incorporation rate variation against deposition temperature (see Fig. 3) which clearly indicates that from deposition temperature of 700°C, P atoms incorporation is somewhat slown down at the expense of a better Si atoms incorporation.

3.4. Sheet resistivity

In Fig. 4 are represented RTLPCVD poly-Si films sheet resistivity variations as a function of deposition temperature before and after an additional RTO treatment (1000°C/20 s) under pure O$_2$ atmosphere. This later process permits to both activate the phosphorus dopant and forms an ultrathin polyoxide which blocks dopant outdiffusion [16]. It can be observed that resistivity starts to strongly decrease (1 order of magnitude) and thereafter saturates at deposition temperature from about 700°C. This trend is probably linked to the dopant segregation at grain boundaries which reduces electrically active dopant atoms in poly-Si grains and especially at lower deposition temperature [17]. In contrast, after the RTO treatment, the resistivity increases with the deposition temperature. A significant variation (two orders of magnitude) is observed for in-

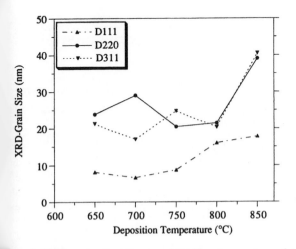

Fig. 2. XRD grain size, D_{hkl}, vs. deposition temperature of RTLPCVD in situ P-doped poly-Si films.

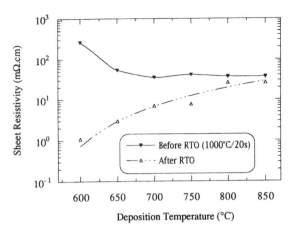

Fig. 4. Sheet resistivity variations as a function of deposition temperature of RTLPCVD poly-Si films before and after subsequent RTO treatment (1000°C/20 s).

itially amorphous layers deposited at 600°C resulting both on the higher dopant concentration and an eventual grain size growth. Furthermore, it might be specified that sheet resistivities of poly-Si films deposited on virgin Si substrates are comparatively lower. In addition, the subsequent RTO treatment (1000°C/20 s) is relatively more beneficial owing to structural changes essentially at the poly-Si/Si interface. Finally, in situ P-doped RTLPCVD poly-Si films with sheet resistivity in the mΩ cm range are routinely obtained when deposition is performed at 750°C.

4. Conclusions

In situ P-doped RTLPCVD poly-Si films were prepared using decomposition reactions of diluted silane ($SiH_4/Ar = 10\%$) with PH_3 (200 ppm) mixture at a fixed total process pressure of 2 mbar. It has been shown that, like for classical LPCVD deposition, adjunction of PH_3 in the Si gas source induced a poly-Si deposition rate reduction. It appeared also that P concentration is principally controlled by deposition temperature and the structural nature of the deposited films. XRD analysis showed that amorphous to crystalline temperature transition was around 650°C. In addition, poly-Si thin films displayed a small-grain crystalline structure without a marked preferential grain orientation. Furthermore, dopant activation process performed sequentially by RTO at 1000°C in pure O_2 atmosphere seems to be operational in order to both activates the phosphorus dopant and forms an ultrathin polyoxide which blocks dopant outdiffusion. Finally, in situ P-doped RTLPCVD poly-Si films with sheet resistivity in the mΩ cm range were routinely achieved at deposition temperature of 750°C followed by optional RTO heat treatment.

References

[1] Wolstenholme GR, Jorgensen N, Ashburn P, Booker GR. J Appl Phys 1987;61(1):225.
[2] Castener LM, Ashburn P, Wolstenholme GR. IEEE Electron Device Lett 1991;12(1):10.
[3] Keyes EP, Tarr NG. Can J Phys 1989;67:179.
[4] Tarr NG. IEEE Electron Device Lett 1985;6(12):655.
[5] Tarr NG, Thomas RE, Wong SK. In: 11th E.C.P.S.E.C Proc., 1992. p. 434.
[6] French PJ, Van Drieënhuizen BP, Poenar D, Goosen JFL, Mallée R, Saro PM, Wollenbuttel RF. J Microelectromech Syst 1996;5(3):187.
[7] Wu CP, Schnable GL, Lee BW, Stricker R. J Electrochem. Soc 1983;131(1):216.
[8] Lemiti M, Semmache B, Lê QN, Barbier D, Laugier A. In: 1st WCPEC Proc., II, 1994. p. 1375.
[9] Harbeke G, Krausbauer L, Steigmeier EF, Widmer AE, Kappert HF, Neugebauer G. J Electrochem Soc 1984;131(3):675.
[10] Mulder JGM, Eppenga P, Hendriks M, Tong JE. J Electrochem Soc 1990;137(1):273.
[11] Learn AJ, Foster DW. J Appl Phys 1987;61(5):1898.
[12] Joubert P, Loisel B, Chouan Y, Haji L. J Electrochem Soc 1987;134(10):2541.
[13] Semmache B, Kleimann P, Le Berre M, Lemiti M, Barbier D, Pinard P. Sensors Actuators A 1995;46-47:75.
[14] Wada Y, Nishamatsu S. J Electrochem Soc 1978;125:1500.
[15] Mackintosh JM. J Electrochem Soc 1961;109:392.
[16] Kallel S, et al., Paper presented in this conference.
[17] Mandurah MM, Krishna, Saraswat C, Robert Helms C, Kamins TI. J Appl Phys 1980;51:5755.

PERGAMON

Materials Science in Semiconductor Processing 1 (1998) 303–315

MATERIALS
SCIENCE IN
SEMICONDUCTOR
PROCESSING

Rapid thermal magnetic annealing as an emerging technology in field-annealing of thin magnetic films for recording heads

F. Roozeboom [a,*], S. Abedrabbo [b], N.M. Ravindra [b], H. Walk [c], M. Falter [c]

[a] *Philips Research, Professor Holstlaan 4, 5656 AA Eindhoven, Netherlands*
[b] *Department of Physics, New Jersey Institute of Technology, Newark, NJ 07102, USA*
[c] *Steag-AST Elektronik, Daimlerstrasse 10, 89160 Dornstadt, Germany*

Abstract

An emerging field where rapid thermal processing (RTP) is now rapidly finding its first acceptance is in the industrial manufacturing of thin-film head devices for magnetic recording. Here soft-magnetic thin-film flux guide structures (usually composed of high-moment alloys containing iron, etc.) are applied onto ceramic substrate wafers (such as Al_2O_3–TiC) of sizes up to 150 mm and subsequently 'activated' by heating and cooling in a magnetic field.

We assessed the advantages of rapid thermal magnetic annealing (RTMA) in a new prototype reactor with an external electromagnet, capable of generating an extremely homogeneous magnetic field of 660 Oe (52.8 kA/m) with field lines parallel across the entire wafer area (150 mm in diameter). Samples with 1 µm thick amorphous iron-alloy layers ($Fe_{77}Nb_{11}N_{10}Si_2$) sputter-deposited onto ceramic substrates of single-crystalline GGG-garnet ($Gd_3Ga_5O_{12}$) were conventionally annealed and RTMA-annealed in N_2/H_2 at temperatures between 550 and 700°C. Structural analysis by transmission electron microscopy (TEM) and electron diffraction showed that the enhanced performance of the RTMA-annealed layers is due to the different nanocrystallization kinetics induced by the fast heating and cooling rates of RTMA.

The ceramic substrate materials normally used in head manufacturing (such as Al_2O_3–TiC) have favorable grey-body properties with high emissivity (≥ 0.7) over a wide range of temperatures (25–700°C) and wavelengths (1.5–10 µm), which excludes the difficulties encountered in pyrometric temperature control of infrared-transparent substrates such as silicon. We conclude that RTMA yields superior soft-magnetic materials, where throughput numbers of ≥ 30 wafers/h are possible. © 1999 Elsevier Science Ltd. All rights reserved.

1. Introduction

One high-technology area where rapid thermal processing is becoming an emerging technology is in the magnetic data storage industry [1]. For the past four decades this industry has seen an above-average annual growth, as drastic as in semiconductor industry. In 1998 the worldwide production of rigid disk-drive heads will exceed for the first time 1 billion pieces[1].

---

* Corresponding author.
[1] Source: Peripheral Research Corp., 1998.

Since the introduction of the magneto-resistive head in 1991 the areal density of commercial hard disk drives increases at a rate of 60% per year [2]. Current areal bit densities in disk-drives are around 1 Gbit/inch^2 and in the laboratories 13 Gbits/inch^2 has been demonstrated so far. The industry has stated objectives of 10 Gbits/inch^2 by the year 2001 and 100 Gbits/inch^2 by the year 2006 [3]. This trend to ultralarge storage densities (i.e. smaller bit lengths and narrower track widths) on recording media requires an increasing medium coercive force, which will prevent the degradation of bits by demagnetization due to the decreasing distance between bit poles and thermal agitation.

This poses in its turn extra requirements on the soft-magnetic fluxguide materials in future recording heads.

In this paper, which is a sequel of earlier studies [4–6], we show that rapid thermal magnetic annealing (RTMA) yields superior materials. After a short discussion on the basic principles of magnetic recording and head structures, we will describe the requirements on future soft-magnetic materials. Next, we will describe the RTMA-reactor we prototyped and we will compare the magnetic performance of thin nanocrystalline iron alloy films (Fe–Nb–Si–N layers) annealed by RTMA with that of conventional furnace-annealed layers. The next section is on the structural analysis of the films by temperature-programmed resistance measurements and TEM and selective area electron diffraction. We conclude with a section on temperature and wavelength dependent emissivity measurements on Al_2O_3–TiC, a commonly used substrate in hard-disk heads and show that this substrate material is an ideal greybody and therefore easier in pyrometric temperature control than silicon.

2. Background on magnetics

2.1. Fundamentals of magnetic recording

The recording and retrieval processes in longitudinal magnetic recording are illustrated for a single track in Fig. 1. In its most simple form the transducer, or recording head, is a ring-shaped electromagnet. It has a soft-magnetic core with a gap facing the magnetic coating on a disk, or in contact with it in case of a tape.

A recording ('write') current causes a fringing field in the gap (see Fig. 1a). Part of this fringe field protrudes outside the gap and (re)directs the magnetization in the medium as it moves along the head. A reversal of the sign of the recording current causes a magnetization reversal or *flux reversal*. The hard-magnetic coating or film retains such an induced magnetization pattern until a new recording field, larger than the coercive field of the medium, is generated to change the direction of the magnetization. During

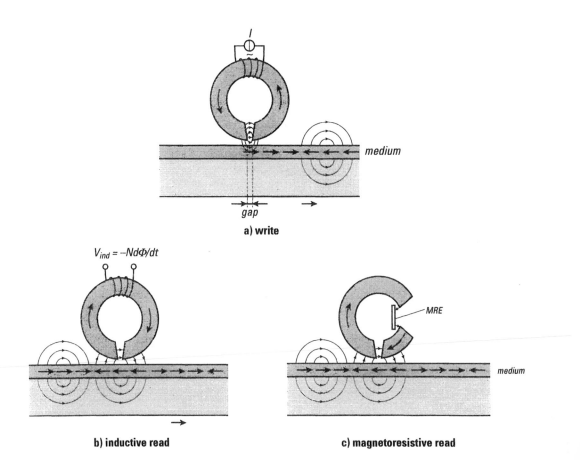

Fig. 1. Principle of inductive write process (a), inductive read process (b) and of magnetoresistive read process (c).

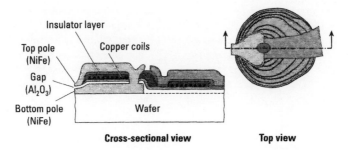

Fig. 2. Side view and top view of a typical thin-film inductive head.

reproduction (play-back or 'reading') the recorded medium moves along the same head (or a read head of similar design). The magnetic transitions in the medium induce magnetic flux changes in the soft-magnetic core. These reversals can be read out either *inductively* (Fig. 1b) or *magneto-resistively* (Fig. 1c).

In the former the flux reversals result in an induced read voltage V_{ind} at the coil terminals according to Faraday's law:

$$V_{ind} = -N \, d\varphi/dt, \tag{1}$$

where N is the number of turns of the coil and $d\varphi/dt$ the flux change. An alternative to inductive reading is magnetoresistive reading, as illustrated in Fig. 1(c). The ring-shaped yoke of the head is interrupted by another gap facing a magnetoresistive element (MRE), composed of a material, currently a 15 nm $Ni_{81}Fe_{19}$ layer [1]; in the future a giant magneto-resistive (GMR) layer stack [7] that changes its electrical resistance upon a change in its magnetization direction due to the magnetic field it senses from the medium. Magneto-resistive reading is much more sensitive, and thus compact, than inductive reading and, unlike inductive reading, it is also fully independent of the

head-to-disk velocity. Today magnetoresistive reading predominates in HDD read heads. Fig. 2 and 3 illustrate how an inductive head and a dual read-write head with magnetoresistive reading are made in thin-film technology.

The static magnetic behaviour of ferromagnetic materials is given by the so-called *B/H loop*, as shown in Fig. 4. Such a curve shows the response of a ferromagnetic material upon the application of an external magnetic field. The magnetic induction B increases upon increasing the field H until a certain maximum, called the saturation induction B_s is reached. If we decrease the field, the induction will also decrease. At $H = 0$ a certain induction remains, called the remanent induction B_r. If we apply a negative field a certain field is needed to reduce the induction back to zero: the coercive field H_c. At larger negative fields saturation is reached again, but in the opposite direction. Thus we traverse a complete hysteresis loop or *B/H* loop.

A recording head is composed of *soft-magnetic* material (Fig. 4a). These are used as *fluxguide* layers in a head because they have a small coercive field H_c and a high *relative permeability* μ_r. For soft-magnetic materials the permeability can be taken from the slope of its B/H loop close to H_c. Due to the low value of H_c, any reversals of the field are converted in reversals of the magnetization in the gap. This causes a field in the *hard-magnetic* recording medium that is large enough to change its magnetization (Fig. 4b). Such media are characterized by a square loop with large H_c and high remanence M_r, so that written magnetizations remain unperturbed and detectable for years. It is obvious that the magnitude of the saturation flux density B_s in the head should be large enough to generate a write field larger than the coercive field H_c of the recording medium.

2.2. Requirements on future soft-magnetic fluxguide materials

Increasing bit density requires improved or new high-coercivity magnetic storage media, since the bits reach their thermal stability limits. Upon further size

**Magnetoresistive-read/
Inductive-write head**

Trailing
write pole

Read gap

Write gap

Disk

Center shield/
Leading write pole

Air bearing surface view

Write pole width

MR sensor

Fig. 3. Side view and top view of a typical dual thin-film head with inductive write and magnetoresistive read part. In the drawing the disk moves to the right.

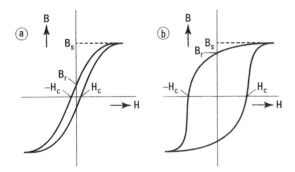

Fig. 4. B/H or hysteresis loops for (a) a soft-magnetic transducer material and (b) a hard-magnetic recording medium.

reduction the anisotropy energy of the bits approaches their thermal energy, which leads to demagnetization from neighboring bits with opposite sign. This trend can be offset by developing magnetic media with higher coercivity, H_c.

However, this poses in its turn extra requirements on the soft-magnetic fluxguide materials in future recording heads. With too high medium coercivity the heads are no longer able to generate the magnetic field necessary to write (H_{write}) on the media without saturating the head material. This field is primarily determined by the saturation magnetization M_s of the soft-magnetic material as:

$$H_{write} = (2/\pi)M_s \arctan(g/2y),\tag{2}$$

where g is the gap length in the head and y is the distance from the medium to the head surface [8].

Usually the saturation magnetic induction B_s ($= \mu_0 M_s$) is used instead of M_s to characterize the limit for a certain material. One can state as a rule of thumb that saturation effects in head materials start as non-linearities at 50 to 80% of the B_s of the material. The need for increasing saturation flux densities has caused recording heads to evolve from monolithic ferrite ($B_s = 0.5$ T) to thin-film heads using permalloy ($Ni_{81}Fe_{19}$, $B_s = 0.9$ T). Future heads may contain amorphous cobalt alloys ($B_s = 1.0$ T), sendust ($Fe_{85}Si_{10}Al_5$, $B_s = 1.0$ T) to nanocrystalline iron alloys ($B_s = 1.5$ T).

High-density magnetic recording is not only characterized by decreasing bit sizes, but also by increasing data rates. These require a high switching speed upon magnetic flux reversal. The main parameter reflecting the switching capability of a soft-magnetic flux guide material is its *relative permeability* μ_r. This parameter should be sufficiently high at high frequencies (e.g. 10–100 MHz range). If the detector element of the read head should efficiently sense the stray flux of the medium, this flux must travel through the ferromagnetic material instead of crossing the non-magnetic gap. It

can be derived that modern heads read only efficiently for values of $\mu_r \geq 3000$ [6].

The relative permeability of a material is not an independent parameter: it is determined by the *magnetic anisotropy energy* which is composed by material-related parameters such as *magnetocrystalline* and *magnetoelastic anisotropy* and by *field-induced anisotropy*. In addition, *shape anisotropy* plays a role during the formation of domain patterns in the poles (see Section 2.3).

2.3. Domain patterns, mechanisms of flux reversal and relative permeability

A thin-film head comprises soft-magnetic materials, which are usually deposited as *flux guide* layers of the order of 1 μm thickness on ceramic wafers. These layers are mostly 'activated' by a furnace anneal in a magnetic field.

In order to be sure that the fluxguide layers are magnetized in a direction perpendicular to the flux path a small non-zero uniaxial anisotropy is required. Usually *magnetic annealing* is used to induce such anisotropy (field-induced anisotropy). The easy axis can be manipulated by the direction of the field during annealing and the magnitude is generally not very large so that high permeabilities can be obtained. The domain pattern has a strong influence on the permeability, as is described in this section.

There are two different mechanisms contributing to a material's permeability (see Fig. 5). In the first case the (induced) magnetization M is parallel to the flux transport direction in the fluxguide, i.e. parallel to the applied magnetic field (along the so-called easy-axis direction). In this case the magnetization changes originate from *domain wall displacement* (see Fig. 5a). They are concentrated in a small area along the domain walls. Due to the limited mobility of the walls this is a relatively slow mechanism yielding very low permeability at high frequencies. Moreover, wall displacement is discontinuous, thus causing Barkhausen noise, and is only effective up to moderate frequencies (10 to 100 kHz).

If a flux reversal is perpendicular to the induced magnetization direction (along the hard axis), these changes occur through *rotation* of the magnetization vector (see Fig. 5b). In contrast to the former mechanism the rotation magnetization process occurs uniformly throughout the fluxguide material. Hence, the local flux density changes per unit time and thus the generated eddy currents and the resulting high-frequency eddy current damping phenomena are considerably smaller than those for the wall displacement mechanism. As a result the rotational magnetization mechanism is effective in flux transport at much higher frequencies (1 to 10 MHz range), depending on the res-

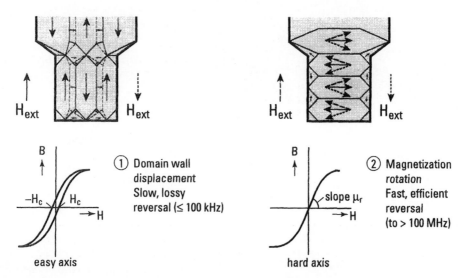

Fig. 5. Flux reversal by domain wall displacement and by rotation of the magnetization. Here an illustration is given for the pole part of a thin film head (cf. Fig. 2).

istivity and the thickness of the fluxguide. It will be clear that for recording applications at these frequencies a domain structure with primarily *rotational* permeability should be induced.

Besides permeability, another important parameter of the magnetic thin films is their *saturation magnetostriction constant,* λ_s [9]. A non-zero magnetostriction in combination with a practically unavoidable stress level in a recording head will often give rise to an excessive stress anisotropy, thus lowering the permeability. It has been reported for permalloy in film heads that even a small change of λ_s from -8×10^{-7} to 5×10^{-7} yields a drastic change in domain configuration [10]. It was also concluded that negative λ_s of the pole material provided more stable heads yielding wiggle-free output [11].

2.4. Flux guide materials.

One class of soft-magnetic materials, known as nanocrystalline materials and having Fe as the main constituent, was found recently to combine a high saturation magnetization (B_s of 1.5–1.7 T) with good soft-magnetic properties ($\mu_r > 3000$, $H_c < 10$ A/m) [12, 13]. These alloys can be written as Fe–TM–N (TM is a group IVa–Va transition metal such as Zr, Hf, Nb or Ta; N is nitrogen). Alloys of Fe–TM–N are very attractive, because they can be easily made by

sputtering from alloy targets in an Ar/N_2-plasma. [14–16].

The as-deposited alloys are amorphous and metastable: they tend to crystallize during the necessary magnetic annealing into a nanocrystalline structure. By the addition of TM and nitrogen the grain size of α–Fe (= body centered cubic, or bcc Fe) is kept below the width of a domain wall, which is of the order of 35 nm for Fe [12, 13]. In this way a low coercivity and a high permeability can be obtained. Research on soft-magnetic materials focuses on nanocrystalline iron films, in which the segregation of TM–N as a second phase (e.g. exothermic, interstitial compounds like face-centered cubic, or fcc NbN) at the triple points and grain boundaries inhibits the further grain growth of Fe (see Fig. 6).

The purpose of this study is to investigate the feasibility of RTMA in a magnetic field to obtain better magnetic properties. Below we will first describe the reactor. Next we will compare the magnetic performance of nanocrystalline $Fe_{77}Nb_{11}Si_2N_{10}$ and its structure to the same materials annealed in a conventional annealing oven. The addition of Si which is highly soluble in Fe serves to promote the nucleation of α–Fe.

3. Experimental

3.1. Reactor design

We co-designed the AST SHS1000 RTMA prototype reactor, which has now been improved to a commercial RTMA version[2]. Our prototype is a 150 mm

---

[2] Commercialized by Steag-AST Elektronik as the AST SHS-ME series, with a 2 kOe electromagnet.

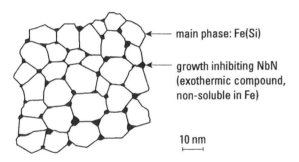

main phase: Fe(Si)

growth inhibiting NbN
(exothermic compound,
non-soluble in Fe)

10 nm

Fig. 6. Schematic view of a nanocrystalline iron film.

wafer, quartz tube reactor in a reflective metal housing, as used in regular silicon processing [17,18]. Fig. 7 shows a schematic drawing of the reactor. The metal housing is surrounded by a computer-controlled electromagnet with fields controllable up to 660 Oe. The directional homogeneity of the field is excellent over the full 150 mm wafer area, as shown in Fig. 8, which shows fully parallel field lines. Also, the variation of the magnitude of the field at maximum value of 660 Oe was measured to be less than 1 Oe across the entire wafer area. This fully computer-automated reactor has also the possibility of wafer rotation in the magnetic field, up to 120 rpm, and of rotating a wafer by 90°. More details can be found elsewhere [4–6].

It will be obvious that, besides the magnetic field homogeneity, the homogeneity of the temperature is important. This should be within $\pm 5°C$ variation across the entire wafer area. One of the reasons is that, in mass-manufacturing thin-film heads with equal performance (signal-to-noise ratio), any head die should contain layers with minimum and equal saturation magnetostriction constant. Thus the requirement becomes:

$$|\lambda_s| \leq 2 \times 10^{-7} \tag{3}$$

For furnace and rapid thermal annealing of many nanocrystalline iron-alloys it turns out that the saturation magnetostriction varies strongly with temperature T [19]:

$$d(\lambda_s)/dT = -4 \times 10^{-8}/K \tag{4}$$

Thus one can estimate the temperature homogeneity specification from the requirements on the magnetostriction to be within $5°C$ across the entire wafer area of 150 mm diameter.

3.2. Materials and thermal processing

Most of the RTMA experiments were performed on $Fe_{77}Nb_{11}Si_2N_{10}$-films of 1 μm thickness. These films were RF-diode-sputtered on substrates of single-crystalline GGG-garnet $(Gd_3Ga_5O_{12})$ measuring 34×5 mm^2 of 0.3 mm thickness. GGG was chosen because its thermal expansion coefficient matches with that of nanocrystalline iron (order 10×10^{-6} °C). Samples for TEM analysis and selected area electron diffraction (SAED) were deposited similarly with 50 nm thickness on ultrathin Si_3N_4-membranes [20].

Samples were annealed stationary in the RTMA reactor using an N_2/H_2 flow of 2.3 l/min. Samples were placed on a 150 mm diameter graphite susceptor. The uniform magnetic field of 660 Oe (52.8 kA/m) was applied in the transverse direction of the strip. The anneals were preceded by a 10 s purge in flowing nitrogen (10 l/min). Below 250–300°C the heating was done in an open-loop control. In our series the ramp rate was always 10°C/s.

However, ramp rates of at least 100°C/s were proven possible without any overshoot above process tempera-

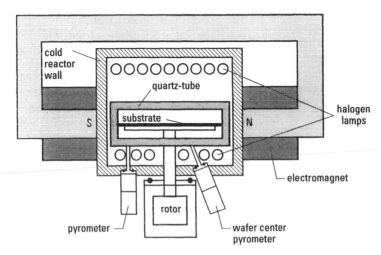

cold reactor wall

quartz-tube

substrate

S   N

rotor

pyrometer

halogen lamps

electromagnet

wafer center pyrometer

Fig. 7. Scheme (not to scale) of the 150 mm wafer diameter prototype RTMA reactor of Steag-AST Elektronik for annealing magnetic thin films in a magnetic field (after Roozeboom et al. [6]).

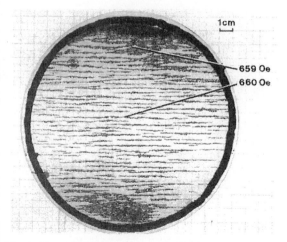

1cm

659 Oe
660 Oe

Fig. 8. Magnetic field lines in the RTMA prototype reactor, visualized by fine magnetic powder sealed in a double glass container in the reactor. The small circle (top left) is a sealed filling hole.

ture. Conventional annealing was done in a vacuum oven [21].

One annealing experiment was done on a commercial 4.5×4.5 inch^2 Alsimag (Al$_2$O$_3$–TiC) wafer with standard thickness of 2 mm with a 1 μm layer of sendust (Fe$_{85}$Si$_{10}$Al$_5$) sputter deposited on top. The processing cycle was identical to the one of the small samples annealed on a graphite susceptor.

3.3. Magnetic performance and structural analysis.

The complex permeability (frequency range 10–13,000 kHz) and the saturation magnetostriction constant were measured as described earlier [19].

The crystallization mechanism was first studied by measuring the sheet resistance of as-deposited 1 μm thick samples by a standard four-point probe technique in vacuum as a function of temperature. Samples were heated to 600°C with a rate of 1°C/min. Next, the crystallization mechanism was also studied by TEM and SAED inspection in a Philips EM400 (120 kV) microscope on as-deposited samples and, for some of the as-deposited samples to determine the crystallization mechanism in situ as a function of temperature.

3.4. Spectral emissivity of ceramic Al$_2$O$_3$–TiC substrate material

The emissivity of these wafer materials was measured in the temperature range of room temperature to 600–700°C, using Fourier transform infrared (FTIR) spectroscopy in a wide spectral range of 6500–500 cm^{-1} (or 1.5–20 μm). The design and the construction of the FTIR apparatus, enabling the determi-

nation of the emissivity at elevated temperatures by measurement of the reflectivity $\rho(\lambda, T)$ and transmissivity $\tau(\lambda, T)$, is described elsewhere [22]. At normal incidence the emissivity $\varepsilon(\lambda, T)$ of a plane-parallel sample is given by:

$$\varepsilon(\lambda, T) = [1 - \rho(\lambda, T)][1 - \tau(\lambda, T)]/[1 - \rho(\lambda, T)$$

$$\cdot \tau(\lambda, T)] \tag{5}$$

For a perfectly opaque body, $\tau(\lambda, T) = 0$, which reduces Kirchhoff's law to

$$\varepsilon(\lambda, T) = 1 - \alpha(\lambda, T), \tag{6}$$

where $\alpha(\lambda, T)$ is the absorptivity of the greybody.

4. Results and discussion

4.1. Magnetic performance.

Fig. 9 shows the relative permeability and the saturation magnetostriction constant for samples with 1 μm Fe$_{77}$Nb$_{11}$Si$_2$N$_{10}$-films on GGG, annealed at different times and temperature. The relative permeability is high after RTMA up to 700°C. This is due to the rapid heating and, in particular, the rapid cooling character of RTMA. Recently, also Kim et al. could relate the better effective permeability of nanocrystalline iron to a faster cooling rate [23].

Above 700°C the permeability decreases steeply. TEM revealed that this is due to growth of Fe crystallites with diameters exceeding 35 nm. A magnetostriction constant λ_s close to zero is obtained at temperatures around 700°C, which is the upper limit to obtain a sufficiently high μ_r. The decrease of λ_s at elevated temperatures is due to the formation of α–Fe, which has a negative λ_s; in contrast, the amorphous FeNbSiN metal glass has a high, positive λ_s.

Compared to RTMA in a static field, conventional vacuum furnace annealing showed a lower permeability of about 7000. Optimum values were obtained for a 2 h anneal at 585°C, see Fig. 9.

4.2. Structural analysis.

The microstructure after optimum conventional annealing (a 2 h anneal at 585°C) and RTMA (30 s/700°C) has been studied. Fig. 10 shows plan-view TEM images together with the corresponding SAED patterns. Both films are polycrystalline. Analysis of the SAED ring patterns provides the lattice spacings and intensities as given in Table 1. The differences for RTMA and furnace annealing are very subtle. The films consist of two phases: α (or bcc)-Fe(Si) and fcc NbN (rocksalt structure). The α-Fe grains are the lar-

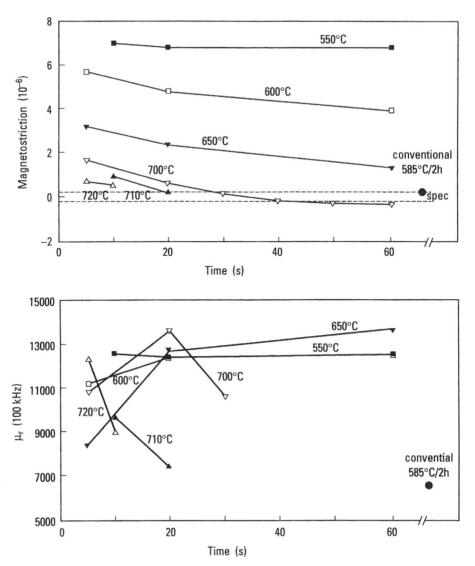

Fig. 9. Saturation magnetostriction coefficient and relative permeability μ_r of 1 µm $Fe_{77}Nb_{11}Si_2N_{10}$-films on GGG at 100 kHz after RTMA and furnace annealing in a magnetic field.

gest fraction. Comparison of the two samples reveals that the grain size of the RTMA sample is slightly smaller. This is seen in the TEM image and confirmed by the SAED patterns: in the rings of the furnace annealed sample more clearly individual spots can be distinguished.

Fig. 11 shows a high resolution TEM micrograph of the RTMA annealed film. One can clearly see the bcc Fe grains with (110) orientation and fcc NbN grains with (111) and (200) orientation (cf. Table 1).

Fig. 12 shows the sheet resistance of an FeNbSiN sample as a function of temperature. Two phase transitions can clearly be distinguished: one at 350°C and one at 530°C. These transitions are irreversible as can be seen from the gradual decrease of the cooling

branch of the curve. Temperature-programmed TEM and SAED confirmed that the first phase transition is the formation of nanocrystalline α-(or bcc) Fe(Si) [(110) and (200) rings] and the second is the segregation of fcc NbN [(111), (200) and (220) rings].

A high relative permeability can only be combined with a low magnetostriction constant by promoting the nucleation and initial growth, whilst inhibiting the further growth of Fe-grains already nucleated and, at the same time, promoting the kinetics of the segregation of NbN along the Fe-grain boundaries and at the triple points. RTMA, with its minimum thermal budget, is an attractive technique to achieve both effects during the (nano)crystallization of FeNbSiN alloy.

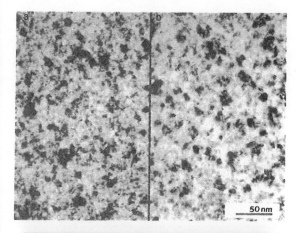

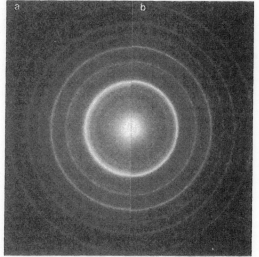

Fig. 10. Plan view TEM and SAED of 50 nm Fe$_{77}$Nb$_{11}$Si$_2$N$_{10}$-films after RTMA for 30 s at 700°C in N$_2$/H$_2$ (a) and conventional furnace annealing (b) for 2 h at 585°C.

This is a general feature, illustrated in Fig. 13. Any reaction, either chemical or solid state (re)crystallization, can be characterized by the Arrhenius equation:

$$r(T) = r_0 \exp[-E_a/RT] \qquad (7)$$

where $r(T)$ is the reaction rate at absolute temperature T, r_0 a constant, R the gas constant and E_a the activation energy (per mol) of the reaction. The activation constant E_a can be calculated from the slope $(-E_a)$ of the line in the plot.

Practically every reaction that is desired is accompanied by one or more undesired side reactions. As stated above the nanocrystallization of α-Fe nuclei starts at 350°C. Literature reports an activation energy of the order of 219 kJ mol^{-1} for this [24]. As soon as the α-Fe nuclei are growing to nanocrystals we would like to inhibit their growth. So the desired reaction is the segregation of the growth inhibitor, NbN. Yet, this reaction starts not until 530°C. By ramping the temperature quickly to this temperature regime one attains this desired reaction regime earlier, thus improving the properties (in our case the relative permeability) of the material being processed. The same holds of course for ramping down when 'freezing' reactions.

This is the virtue of rapid thermal annealing (RTA) in general: compared to conventional furnace one has a stronger tool to promote and control the reaction. Thus RTA can be used better to thermally 'switch' chemical or solid state reactions on and off.

4.3. Wafer characteristics

4.3.1. Materials

The wafers used in thin-film head technology are typically ceramic wafers sized up to 6-inch square or round wafers of typically 2 mm thickness. In disk

Table 1
Lattice spacings (Å) and intensities of 50 nm Fe$_{77}$Nb$_{11}$Si$_2$N$_{10}$-films, RTA and furnace annealed (left two columns). For comparison the calculated lattice spacings of α–Fe and of fcc NbN are added.

RTP	Furnace	α–Fe	fcc NbN
2.459/15	2.441/13	–	2.536/70/(111)
2.173/12	2.130/9	–	2.197/100/(200)
2.036/100	2.016/100	2.027/100/(110)	–
1.508/9	1.495/9	–	1.553/49/(220)
1.434/19	1.420/23	1.433/13/(200)	–
1.284/2	1.281/2	–	1.324/17/(311)
			1.268/13/(222)
1.167/30	1.157/36	1.170/21/(211)	1.098/7/(400)
1.012/5	1.005/7	1.013/6/(220)	1.008/5/(331)
			0.982/13/(420)
0.903/8	0.895/11	0.906/7/(310)	0.897/8/(422)
0.826/2	0.816/2	0.828/1/(222)	0.845/2/(511)
0.763/7	0.757/10	0.766/6/(321)	0.777/2/(440)

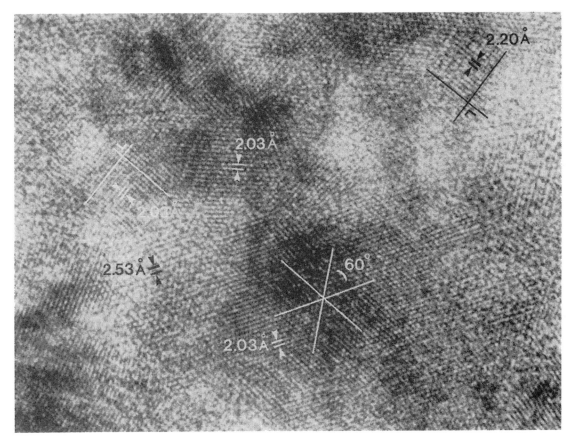

Fig. 11. High resolution TEM micrograph of 50 nm $Fe_{77}Nb_{11}Si_2N_{10}$-films.

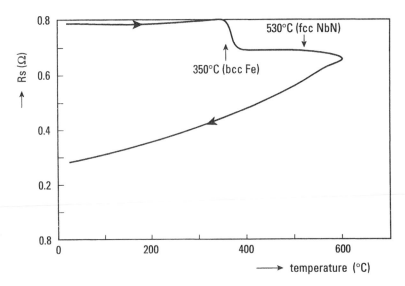

Fig. 12. Sheet resistance of a 1 µm $Fe_{77}Nb_{11}Si_2N_{10}$-film on GGG as a function of temperature. The as-deposited sample was heated to 600°C in vacuum at a rate of 1°C/min and then cooled down.

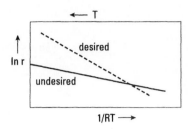

Fig. 13. Arrhenius plot for a thermal process causing a desired reaction and an undesired reaction.

drive heads [1] and also in digital tape recording [25] the layers for the thin-film heads are deposited onto a substrate which must exhibit good wear and corrosion resistance, the right thermal conductivity and expansion coefficient, and good machinability into an air-bearing or tape bearing surface. Preferred materials are Ni–Zn ferrite and Al_2O_3–TiC [26].

In order to get an idea about the feasibility of RTA in a magnetic field for application in thin-film data storage head manufacturing technology, etc., it is interesting to compare the (thermo)physical properties of the substrate materials used with those of silicon. Table 2 lists some of these properties. From Table 2 one can conclude that the ceramic materials have comparable heat capacity, but a lower thermal conductivity. Also the total weight of a wafer with equal diameter will be some 4 times higher than that of a typical Si-wafer. Yet, the emissivity of the ceramics is higher (see below) and the typical processing temperatures are typically in the medium range of 400–700°C. This means that the process cycle times will be of the same order (typically 2–3 min per wafer). The thermal conductivity of the ceramics is somewhat lower, which means that the influence of pattern-induced temperature non-uniformity may cause more problems. Yet, this is compensated by the more 'black' ceramic substrate bodies.

Fig. 14 shows a typical temperature–time cycle for an RTMA process on a 4.5×4.5 inch^2 Alsimag (Al_2O_3–TiC) wafer with standard thickness of 2 mm with a 1 µm layer of sendust ($Fe_{85}Si_{10}Al_5$). Here we used a ramp rate of only 10°C/s. Ramp rates of 50°C/s

were proven possible without overshoot above the set point temperature. It shows indeed that throughputs of 20–30 wafers/h are possible.

Fig. 15 shows the spectral emissivity (in fact the emittance) of a Al_2O_3–TiC wafer of 2.8 mm thickness. The wafer was measured on its polished side. The following conclusions can be drawn: Al_2O_3–TiC has fully opaque ($\tau = 0$) character in the full spectral range of 6500–500 cm^{-1}. Only in the far-infrared region from 1000–500 cm^{-1} it shows a drastic decrease in emissivity. In the 2.7–2.8 µm range (e.g. the pyrometer band of our Steag-AST Elektronik RTMA reactor) the emissivity varies only between 0.7–0.8, thus rendering this wafer material almost to a grey-body. Since all of the thin-film head processing is well within the 25–700°C range, and most pyrometers operate above 1000 cm^{-1} (below 10 µm), we can conclude that the use of pyrometric temperature control is less complex than in the case of semiconductor substrates or, more general, that RTMA of thin magnetic films on ceramic wafers in thin-film head manufacturing is not more complicated than RTP of silicon wafers.

5. Conclusions

This paper focuses on rapid thermal magnetic annealing of soft-magnetic metallic flux guide layers in thin-film recording heads. We designed a prototype RTMA-reactor to anneal these films in the presence of a magnetic field. Using the example of nanocrystalline iron as a soft-magnetic flux guide material it is illustrated that RTMA with its inherent low thermal budget can be used successfully in order to obtain the proper magnetic performance of soft-magnetic materials in high-density magnetic data storage. Structural analysis by TEM and electron diffraction showed that the enhanced performance of the RTMA-annealed layers is due to the different nanocrystallization kinetics induced by the fast heating and cooling rates of RTMA.

The ceramic substrate materials normally used in head manufacturing (such as Al_2O_3–TiC) have favor-

Table 2
Physical parameters of ceramic and silicon substrate materials

Parameter	Al_2O_3–TiC	Ni–Zn ferrite	Silicon
Heat capacity (J/g K)	0.5–1.0	0.5–1.0	0.7
Thermal conductivity (W/m K)			
Room temperature	10–20	3.5–4.0	156
700°C			32
1100°C			24
Specific weight (g/cm^3)	4.2	5.3	2.3
Typical wafer thickness in manufacturing (mm)	2	2	1

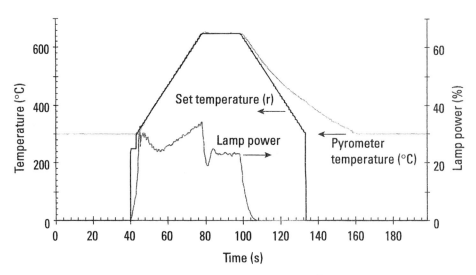

Fig. 14. Typical temperature-time profile for an RTMA process on a 4.5 × 4.5 inch² ceramic (Al$_2$O$_3$–TiC) wafer of 2 mm thickness with a 1 μm layer of sendust (Fe$_{85}$Si$_{10}$Al$_5$).

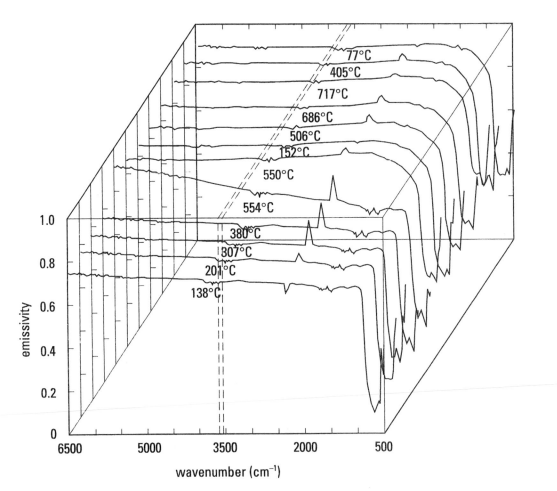

Fig. 15. Emissivity spectrum of an Al$_2$O$_3$–TiC wafer as a function of temperature. The broken lines represent a 2.7–2.8 μm pyrometer pass-band of our RTMA reactor.

able grey-body properties with high emissivity (≥ 0.7) over a wide range of temperatures (25–700°C) and wavelengths (1.5–10 μm), which excludes the difficulties encountered in pyrometric temperature control of infrared-transparent substrates such as silicon. We conclude by stating that RTMA of thin magnetic films on ceramic wafers in thin-film head manufacturing is not more complicated than that of silicon wafers in semiconductor manufacturing.

Acknowledgements

The TEM and SAED analyses by J.J.T.M. Donkers are acknowledged, as well as the temperature dependent resistance measurements by R. van der Rijt.

References

[1] Ashar KG. Magnetic disk drive technology, heads, media, channel, interfaces and integration. New York: IEEE Press, 1997.

[2] Bajorek CH, Mee CD. Data Storage 1994;1(1):23.

[3] Kryder M. Data Storage 1998;5(4):49.

[4] Roozeboom F, Ruigrok JJM, Klaassens W, Kegel H, Falter M, Walk H. Mater Res Soc Symp Proc 1996;429:203.

[5] Roozeboom F. In: Roozeboom F, editor. Advances in rapid thermal and integrated processing. Dordrecht, The Netherlands: Kluwer Academic Publishers, 1996. p. 521–52.

[6] Roozeboom F, Bloemen PJH, Klaassens W, Van de Riet EGJ, Donkers JJTM. Philips J Res 1998;51(1):59–91.

[7] Kools JCS, Coehoorn R, Folkerts W, De Nooijer MC, Somers GHJ. Philips J Res 1998;51(1):125–48.

[8] Ruigrok JJM. Short-wavelength magnetic recording, new methods and analyses. Oxford, UK: Elsevier, 1990. p. 205.

[9] De Wit HJ. Rep Prog Phys 1992;69:113.

[10] Narishige S, Hanazono M, Takagi M, Kuwatsuka S. IEEE Trans Magn MAG 1984;20:848.

[11] Hanazono M, Narishige S, Hara S, Mitsuka K, Kawakami K, Sugita Y. J Appl Phys 1987;61:4157.

[12] De Wit HJ. J Magn Magn Mater 1989;79:167.

[13] Herzer G. IEEE Trans Magn 1990;26:1397.

[14] Nakanishi K, Shimizu O, Yoshida S. J Magn Soc Jpn 1991;15:371.

[15] Nago K, Sakakima H, Ihara K. J Magn Soc Jpn 1991;15:365.

[16] Ishiwata N, Wakabayashi C, Urai H. J Appl Phys 1991;69:5616.

[17] Roozeboom F. In: Fair RB, editor. Rapid thermal processing, science and technology. New York: Academic Press, 1993. p. 349–423.

[18] Deutschmann L, Glowacki F. In: Roozeboom F, editor. Advances in rapid thermal and integrated processing. Dordrecht, The Netherlands: Kluwer Academic Publishers, 1996. p. 431–41.

[19] Roozeboom F, Dirne FWA. J Appl Phys 1995;77:5293.

[20] Jacobs JWM, Verhoeven JFCM. J Microsc 1986;143(1):103.

[21] De Wit HJ, Witmer CHM, Dirne FWA. IEEE Trans Magn 1987;23:2123.

[22] Markham JR, Kinsella K, Carangelo RM, Brouilette CR, Carangelo MD, Best PE, Solomon P. Rev Sci Instrum 1993;64:2515.

[23] Kim KY, Lee JS, Noh TH, Kang IK, Kang T. J Appl Phys 1994;75:6943.

[24] Al-Haj M, Barry J. J Mater Sci Lett 1997;16:1640.

[25] Zieren V, Somers G, Ruigrok J, De Jongh M, Van Straalen A, Folkerts W, Draaisma E, Pronk F, Mitchell T. IEEE Trans Magn 1993;29:3064.

[26] Mee CD, Daniel ED. Magnetic recording, vol. I: Technology. New York: McGraw-Hill, 1987.

PERGAMON

Materials Science in Semiconductor Processing 1 (1998) 317–323

MATERIALS
SCIENCE IN
SEMICONDUCTOR
PROCESSING

Epitaxial growth of SiGe layers for BiCMOS applications

J.L. Regolini [a,*], J. Pejnefors [b], T. Baffert [a], C. Morin [a], P. Ribot [a], S. Jouan [a], M. Marty [c], A. Chantre [a]

[a]*FRANCE TELECOM-CNET Grenoble, 28 Ch. Du Vieux Chêne, 38243 Meylan Cedex, France*
[b]*Department of Electronics, Royal Institute of Technology, EKT, SE 164 40 Kista, Sweden*
[c]*SGS-Thomson, BP 16, 38921 Crolles Cedex, France*

Abstract

Silicon–germanium materials have introduced the opportunity to engineer the energy band gap of Si, leading to a wide range of microelectronic device applications. The growth of high quality SiGe layers by chemical vapor deposition at reduced temperature and pressure has already been reported from our GRESSI Program [1]. In addition to these results we have studied the thermal stability of Si–Ge strained layers to be used as the base of heterojunction bipolar transistors (HBTs) or the channel of MOS structures. Indeed, layer growth conditions such as temperature, Ge content and selectivity/non-selectivity may induce structural defects such as misfit dislocations, leading to leaky devices and less performant ICs. Using an industrially available CVD single wafer reactor we have fabricated and studied Si/SiGe stacks for CMOS and BiCMOS applications. In the present paper the obtained electrical results on HBTs are correlated with the growth parameters and observed structural defects, in order to optimise device characteristics and process windows. © 1999 Elsevier Science Ltd. All rights reserved.

Keywords: Heterojunction bipolar transistor; SiGe epitaxy; Misfit dislocations; Critical layer thickness; Single wafer epitaxy; BiCMOS process

1. Introduction

SiGe has been studied for many years as a material to be used in Si ultra large scale integration (ULSI) microelectronics. The performance of Si has been extended through higher speed devices using the heterojunction bipolar transistors (HBT) and better current drivability in CMOS devices. Maybe the most successful device is the HBT with several impressive results on transit frequency up to 160 GHz [2]. However, most of these extreme results are far from industrial manufacturability, due to a lack of best performances in other parameters and considerable increase in process complexity. More realistic figures are in the range of 20 to 50 GHz using, for example, a standard 0.5 μm BiCMOS process [3].

Device integration as well as cost are some of the most important issues for the success of SiGe. Thus, after a long period of material studies and device feasibility demonstrations, the time is coming for manufacturability of HBT integration and commercial developments as recently reported [4].

Since the early days, and up to now, SiGe growth has been carried out under ultra high vacuum (UHV) conditions by molecular beam epitaxy (MBE) or chemical vapor deposition (CVD). More recently, other pressure ranges have been introduced, going from reduced pressure (RP) to atmospheric pressure (AP) with advantages and disadvantages for each of these techniques. At any working pressure, the temperature range is the so called low temperature range, from 500 to 800°C, thus sample surface preparation is a major issue and a function of the used equipment.

* Corresponding author. Tel.: 33-476-764-247; fax: +33-476-903-443; e-mail: jorgeluis.regolini@cnet.francetelecom.fr

1369-8001/99/$ - see front matter © 1999 Elsevier Science Ltd. All rights reserved.
PII: S1369-8001(98)00025-0

The fabricated heterostructures of interest are meta-stable. Si and Ge are miscible over the entire compositional range, but a maximum of 4% lattice parameter mismatch leads to relaxation through the generation of misfit dislocations during growth or during elevated temperature treatments. To maintain the transport advantages, the epitaxial SiGe layers should be commensurate with the substrate, i.e. strained. Misfit dislocations may generate threading components, which propagate through the epitaxial layer inducing leakage currents and consequently device performance degradation.

Wafer surface preparation and growth parameters may play an important role on defect formation. For example: using a commercial APCVD reactor, Burghartz et al. [5] observed the formation of misfit dislocations in the 600/700°C growth temperature range and about 20% Ge fraction (selective epitaxy, box type Ge profile). They attributed the defects to the high Ge content for that HBT base thickness, i.e. a problem related to the SiGe equilibrium critical thickness [6]. On the other hand, using a higher temperature range (700/750°C), RP commercial reactor and selective epitaxy with up to 20% Ge content and equivalent layer thickness, Vook et al. [7] reported no misfit dislocation formation in a graded Ge profile. These two examples show that the same average Ge concentration may produce different results according to the experimental conditions.

As we will see later, for a given Ge average concentration, the manufacturing parameters such as surface cleaning and heterogeneous defect nucleation centers may give different crystallographic qualities. In the present work, we try to correlate what we call the manufacturing parameters with the generation of misfit dislocations of an integrated BiCMOS technology under development.

1.1. Base technologies and experimental details

Two different BiCMOS support technologies have been used to introduce the SiGe base HBT with a minimum perturbation for the whole process. In the first case, we have used a selective epitaxy to replace the existing ion implanted base, while in the second a non-selective epitaxial process essentially to contact the base over the field oxide through the poly-SiGe layer, thus minimizing the base/collector capacitance. According to the literature, these two approaches have shown equivalent performances and the choice is finally a matter of the support technology and equipment availability. Selectivity has the advantage of directly replacing the existing implanted base, while the non-selectivity is more independent of pattern filling ratio.

1.1.1. 0.5 μm BiCMOS process

This process has already been presented [8] and will not be described here in detail. Essentially a poly buffered LOCOS (PBL) delimits the active regions composed of CMOS and bipolar devices. After full CMOS processing, the bipolar base regions are opened by chemical etching using diluted HF and rinse. For the selective epitaxial stack, we used an industrial 200 mm single wafer CVD module working at about 10 Torr. A pre-bake step is carried out under H_2 at 900°C during 2 min, followed by the selective epitaxial growth (SEG) using DCS/GeH_4 (10% in H_2)/ $B_2H_6(PH_3)$ and H_2 as a carrier gas. The growth temperature was fixed at 750°C and with a maximum Ge concentration around 10% we do not observe misfit dislocations after chemical decoration and/or TEM. The temperature was chosen to be 750°C for high growth rate and defect free layers. Electrical characteristics and BiCMOS performances have already been presented [8] and will be discussed here only for the sake of comparison with the next more recent process.

1.1.2. 0.35 μm BiCMOS process

This process also uses PBL isolation and a non-selective (NSEG) SiGe epitaxy is grown in a quasi self-aligned emitter/base architecture. Details of this process will be presented elsewhere [9]. Essentially a 'SiGe drift-transistor' profile [5] is employed for thermal budget compatibility with the 0.35 μm CMOS. An a-Si layer is used as a hard mask for wet oxide removal from the base areas prior to the NSEG, thus minimizing loading effects against different mask layouts. For the NSEG, SiH_4 replaces DCS as the Si precursor. The temperature was kept in the 650/700°C range to have equivalent growth rate as in Section 1.1.1. In this NSEG case we have observed (by chemical decoration and TEM) a variable density of misfit dislocations going all the way through the individual active zones, from one side to the opposite side. This particular behavior, observed mainly in big active areas ($> 100 \times 100$ μm^2) and 700°C growth temperature led us to initiate a more in depth study to elucidate defect origins and their secondary effects.

1.2. Material issues

In Fig. 1, we show some of our most used Ge profiles for the HBT base (doping profiles are not presented). Curves (a) and (b) are obtained by NSEG, at 650 and 700°C, respectively, and we observed some misfit dislocations by optical microscopy after chemical decoration. On the other hand, curve (c) is from a SEG at 750°C and we do not observe misfit dislocations despite the higher average Ge content.

When defects are observed, they are already present after growth at 700°C, prior to any process thermal

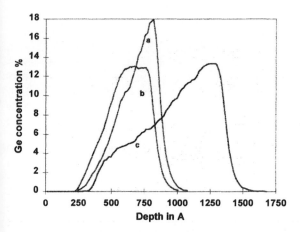

Fig. 1. Three Ge profile (SIMS) types used within this study. (a) and (b) were by NSEG and (c) by SEG. In (a) and (b) we observed misfit dislocations while in (c) we did not.

treatments. The NSEG gives single crystal SiGe into the active open areas and poly-SiGe over the oxide regions. A TEM cross section showing the transition from epi to poly, which is inherently a highly defective region, just above the bird's beak of the PBL is reported in Fig. 2. This region may act as a heterogeneous source for misfit dislocations which propagate according to the ⟨110⟩ directions essentially at the Si (substrate)/SiGe interface.

The origin of misfit dislocations is the strain field generated by heterojunction. Their generation is facilitated by the growth temperature as well as the presence of heterogeneous nucleation centers which may come from different origins, such as surface cleaning, reactive gas contamination and structural defects or the substrate strain field [10].

In fact, as pointed out by Schonenberg et al. [10], there are several potential sources for the generation of line defects in a stressed layer. In Fig. 3, we observe

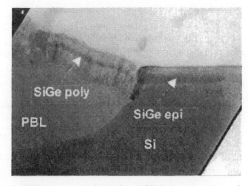

Fig. 2. TEM cross section of a NSEG layer at the epi–poly SiGe transition. This transition region is composed of single crystal SiGe with a high density of twins and stacking faults. Layer continuity is well obtained.

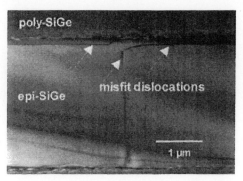

Fig. 3. TEM plan view of an active area in which some dislocations (arrows) have been induced at the transition epi/poly SiGe.

misfit lines originated at the poly-SiGe/epi-SiGe interface. Once formed, they propagate from one side to the opposite side. The only nucleation site is that interface in our NSEG. On the other hand, when dislocations are observed in the SEG case, they are originated inside an active area by some crystallographic defects or particulates.

It has also been shown that the dislocation density is severely reduced in small active areas when compared to adjacent large areas after a selective growth [11]. We do not observe dislocations on layers grown on blanket wafers under the same epitaxial conditions of batch wafer; however, on batch wafers there is an area-dependent defect density which is the focus of the present work.

We used PBL as a wafer support for all the experiments, thus the strain level before epitaxial process is the same for all the wafers. Dislocation counting was carried out after chemical decoration and density measurements were done optically under Nomarski contrast. The total stack thickness was measured with cross-sectional SEM and also SIMS. Ge profiles were obtained by SIMS. The evaluated structures are shown in Fig. 1: they consist in a buffer layer followed by SiGe (constant and graded) and finally a capping layer.

2. Results and discussion

2.1. Misfit dislocation measurements

For the production of misfit dislocations the lattice mismatch stress should exceed some critical value which may be predicted from energetic considerations [12]. In such a way a theoretical equilibrium stability criterion is defined from which there is a critical thickness above which there is crystal relaxation by the generation of misfit dislocations. More recently, Stiffer et al. [13] showed that films thicker

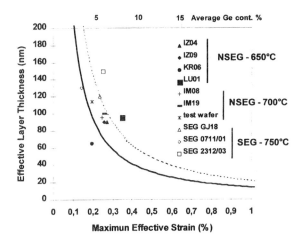

Fig. 4. Effective layer thickness versus maximum effective strain obtained from Ref. [10] for different batch wafers obtained under different epitaxial process conditions.

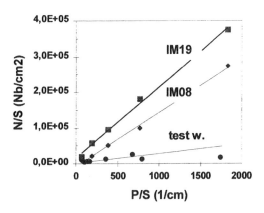

Fig. 5. Misfit dislocation density as a function of P/S for the split of wafers grown at 700°C NSEG from Fig. 4.

than the predicted critical thickness may be grown defect free, bringing some empirical stability criterion, about 1.65 times the theoretical critical thickness. From these considerations we understand that experimental conditions are of crucial importance in obtaining defect free SiGe layers for device applications.

To calculate the effective strain of grown layers, we followed the procedure shown by Tsao et al. [10, 14] for a film of non-uniform composition in which the capping layer is also included into the effective layer thickness. In Fig. 4, we show the theoretical (continuous line) and empirical (dotted line) curves of the effective layer thickness as a function of the maximum effective strain as well as the superposition of values corresponding to several measured layers from different batches and experimental conditions.

The sample points in Fig. 4 are summarized as follow:

1. SEG-750: no defects as dislocation lines have been observed on these wafers with a Ge content corresponding to curve (c) in Fig. 1. Under these growth conditions there is no misfit dislocations on these types of wafers despite the 750°C growth temperature.

2. NSEG-700: generally we observed dislocation lines on large areas ($>100 \times 100$ μm^2) of all these samples. These experimental conditions, i.e. non-selectivity and 700°C, give more instable working points and the electrical results confirm this conclusion.

3. NSEG-650: growth temperature was 650°C and is the set in which we have practically no misfit dislocations under non-selective conditions. In sample LU01 we observe some of them only in large areas.

To better understand the nucleation of dislocations we can describe the total number of dislocations (n_{tot}) in an active area as

$$n_{tot} = n_S \cdot S + n_P \cdot P \tag{1a}$$

where n_S is the surface dislocation source in a number of dislocations per cm^2, S is the surface, n_P is the perimeter dislocation source in a number of dislocations per cm and P is the perimeter. n_S corresponds to dislocations nucleated by substrate dislocations, precipitates and dust particles [11, 15]. n_P is mainly due to the dislocations nucleated at the heavily defective region present in the epi–poly transition (see Figs. 2 and 3). A closer look at the origin of the dislocation nucleation may be given by the dislocation density, defined as the number of dislocations per cm^2 plotted as a function of P/S. The slope represents the dislocations created at the perimeter and the interception with the y-axis represents the dislocations originated at the surface. From Eq. (1a),

$$\frac{n_{tot}}{S} = n_S + n_P \cdot \frac{P}{S} \tag{1b}$$

In Fig. 5, the dislocation density, calculated by taking the average number of dislocations in several identical active areas divided by their surface, is plotted versus P/S for the NSEG samples: test wafer, IM08 and IM19, grown at 700°C and for squares from 20×20 to 600×600 μm^2. An increase in the dislocation density as a function of P/S is noted for all samples showing that the perimeter length is of primary importance for the generation of dislocations. The n_S values are negligible for all three sets of data, i.e. the surface contribution to the total number of defects is very low (though the importance of the starting surface cleaning and epitaxial quality in order to avoid heterogeneous nucleation centers).

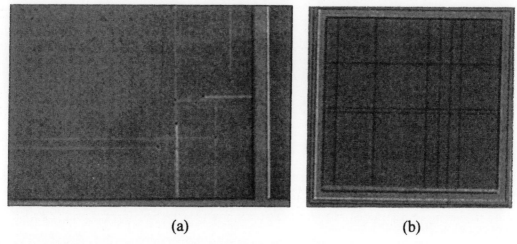

(a)   (b)

Fig. 6. Nomarski optical micrographs showing the influence of surface treatment: (a) SEG of $300 \times 300 \ \mu m^2$ and (b) NSEG of $100 \times 100 \ \mu m^2$. Both wafers under equivalent conditions, thickness and [Ge].

Moreover, a tendency of increasing linear dislocation density with the distance from the theoretical equilibrium curve of Fig. 4 is also observed.

Using Eq. (1a) to describe the number of misfit dislocations in SEG films should result in a small value in n_P, since there is no epi–poly transition and the dominant source for dislocation nucleation is the active area surface. This has been observed by other authors on selective growth [11, 15]. Thus, the largest contribution to n_S for SiGe films grown on Si is the contamination of the Si surface before epitaxy. This has been confirmed by depositing a SEG film with incomplete surface cleaning. In this case dislocations are observed as shown in Fig. 6 which is a Nomarski optical micrograph of the contaminated area.

The dislocations are nucleated inside the active area terminating into another surface defect, dislocation or at the field oxide sidewalls. In earlier work with SEG SiGe, we have observed an increased leakage current in the base/collector junction depending on the pre-epitaxial cleaning procedure [16].

2.2. Thermal stability

The thermal stability of NSEG layers grown at 650°C has been studied within the technologically useful thermal budget, i.e. between 900 and 1000°C. In Table 1 we have a qualitative comparison of different samples from Fig. 4 after thermal cycles under N_2 gas. We measured, only in one case, an increase in misfit dislocations related with the total thermal budget.

For sample LU01, we performed a more quantitative analysis and plotted the dislocation generation rate $(1/cm^2 \ s)$ versus the reciprocal absolute temperature (Fig. 7). Misfit dislocation densities have been measured on several square areas of 600 μm by side and all the thermal cycles have been of 5 min.

Values plotted in Fig. 7 mean that we are in a low dislocation density regime, in the order of $1000/cm^2$ after growth for this worst case (see Fig. 4). Higher densities will give layers not suitable for device applications [15]. From Fig. 7 the calculated activation energy of 3.9 eV which is too high for only misfit dis-

Table 1

	LU(01) metastable	IZ(09)	IZ(04)	KR(06) stable
Reference	some dislocation	no dislocation	a few segments	no dislocation
900°C, 5 min	some dislocation			
950°C, 5 min	+	no dislocation		
1000°C, 1 min	some dislocation			
1000°C, 5 min	+ +	no dislocation	a few segments	no dislocation
1000°C, 30 min	+ + +	no dislocation		

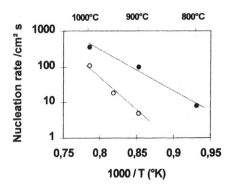

Fig. 7. Misfit dislocation nucleation rate versus inverse anneal temperature for the LU01 sample from Fig. 4. Closed circles are from Ref. [15] with an activation energy of 2.3 eV. Open circles are this work and we calculated 3.9 eV.

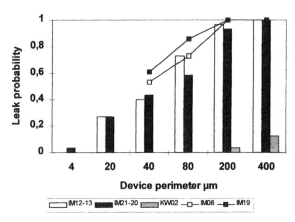

Fig. 8. Leakage current and misfit dislocations for 3 different wafers.

locations nucleation. This obtained value is closer to a plastic strain relaxation, i.e. the extend of misfit relaxation by nucleation and glide [15].

We should point out that all tested samples with maximum effective strain (Fig. 4) below 0.3% are stable under severe thermal treatments up to 1000°C during 30 min.

2.3. Electrical measurements

Electrical measurements have been performed on the HBTs to correlate the observed misfit dislocation and the induced leakage current I_{CEO}. Measurements have been done at $V_{CE} = 2.5$ V and with an open base. We used 30 identical transistors distributed uniformly over the wafer and the 'leaky transistor criterion' was $I > 100$ pA. In Fig. 8, the leakage probability (defined as the number of leaky devices over the total number of tested devices) is plotted versus the perimeter of the emitter window for wafers IM12-13 and IM20-21 grown NSEG at 700°C and corresponding to the same splits as wafers IM08 and IM19, respectively. For comparison, results from wafer KW02, NSEG batch at 650°C are also plotted.

For the batch IM, we observed that all transistors with a perimeter larger than 200 μm are leaky. This result correlates fairly well with our optical observations after epitaxial growth where we observe a number of dislocations in all those areas. From Fig. 8, the probability of having a leaky transistor increases with the perimeter length, for devices with a perimeter smaller than 200 μm. This result is equivalent to the two samples IM08 and IM19 (Fig. 5).

The calculated average leakage current (not shown here) is larger for IM19 which correlates with the higher dislocation density shown in Fig. 5.

In Fig. 8 we also plot the probability of having dislocations in the different active areas from the optically

observed wafers from the same batches. A good correlation is obtained with the electrical results for devices from 10×10 to 100×100 μm². In other words, the same behavior is observed with both methods, but a larger value is found with optical observations, which could be explained by the fact that not all defects are electrically active.

Batch KW grown at 650°C with a thinner SiGe layer but a higher strain shows no dislocations after growth and no transistors leak up to a perimeter of 80 μm while at 400 μm only 12% are leaky.

3. Conclusion

We have grown non-selective $Si/Si_{1-x}Ge_x/Si$-films for HBTs. We have observed that a number of misfit dislocations nucleate during growth, mainly at 700°C. These lines originate at the epi/poly SiGe interface under non-selective conditions. Selective epitaxy gives lower dislocation density for equivalent thickness and Ge contents even at 750°C growth temperature. To decrease (or avoid) these defects, keeping constant the alloy composition and base thickness, it is necessary to decrease the growth temperature (thus, increasing the process time). At 650°C we have obtained 'defect free' stacks and no leaky devices.

At this growth temperature, if the maximum effective strain–effective layer thickness values are within the metastable region the subsequent thermal treatments do not increase the dislocation density. However, far from the empirical stability curve the process thermal cycles will nucleate an important misfit dislocations density. Moreover, we have shown that the number of dislocations per cm² is correlated to the maximum effective thickness and strain and they originate from the periphery of the active areas.

We have briefly correlated the presence of dislocations with electrical measurements and we observed leaky devices in batches with visible dislocations lines after chemical decoration. However, more work has to be done to understand the origin of the dislocations, in principle their appearance is related with the epi–poly transition in the non-selective growth as we do not observe misfit dislocations under selective growth or onto clean blanket wafers under equivalent conditions.

Acknowledgements

The authors are grateful to J. Del Medico for his contribution to epitaxial growth and R. Devine for sample annealing. This work has been carried out within the GRESSI consortium between CEA-LETI and France Telecom-CNET.

References

[1] Regolini JL. In: Riley T, Gelpey J, Roozeboom F, Saito S, editors. Rapid thermal and integrated processing VI, vol. 470. 1997. p. 99.
[2] Schuppen A et al. IEDM Tech Dig 1995;743.
[3] Harame DL et al. IEEE Trans ED 1995;42(3):469.
[4] Holton WC. Solid-State Technol 1997;November:119.
[5] Burghartz JN et al. IEEE 1993 Bip. Circuits and Tech. Meeting. p. 55.
[6] People R. IEEE-QE 1986;22(9):1696.
[7] Vook D et al. IEEE-ED 1994;41(6):1013.
[8] de Berranger E et al. Thin Solid Films 1997;294:250.
[9] Chantre A et al. Bip. Circuits Tech. Meeting, 1998.
[10] Schonenberg K et al. J Mater Res 1997;12:364.
[11] Noble DB et al. Appl Phys Lett 1990;56:51.
[12] Matthews JW et al. J Cryst Growth 1974;27:118.
[13] Stiffler SR et al. J Appl Phys 1992;71:4820.
[14] Tsao JY et al. Appl Phys Lett 1988;53:848.
[15] Houghton DC. J Appl Phys 1991;70:2136.
[16] Assous M et al. J Vac Sci Technol B 1998;B16(3):1740.

PERGAMON

Materials Science in Semiconductor Processing 1 (1998) 325–329

MATERIALS
SCIENCE IN
SEMICONDUCTOR
PROCESSING

Selective doping of silicon by rapid thermal and laser assisted processes

U. Besi-Vetrella [a,*], E. Salza [a], L. Pirozzi [a], S. Noel [b], A. Slaoui [b], J.C. Muller [b]

[a]ENEA-CR Casaccia, sp 064, Via Anguillarese 301, 00060 Rome, Italy
[b]PHASE-CNRS, 23 Rue du Loess, B.P. 20, 67037 Strasbourg Cedex, France

Abstract

The selective doping technique, made by the combination of spin-on dopant (SOD) source deposition, rapid thermal annealing (RTA) and laser treatments is proposed as an innovative process for large area devices, like silicon solar cells.

Rapid thermal diffusion (RTD) is first carried out from phosphorus SOD layers to form a lightly doped junction followed by pulsed laser irradiation to induce overdoping in selectively chosen regions.

Here we present extensive study on the dependence of selective doping efficiency through different working variables, such as dopant source dilution, diffusion temperature and time for RTPs, and power and translation velocity for lasers. Electrical and structural characterizations have been performed by using several techniques: SIMS, stripping-Hall, four-point probe resistivity, SEM and TEM analysis.

The combined use of these processes has been applied to the realization of selective emitter structures for silicon solar cells. © 1999 Elsevier Science Ltd. All rights reserved.

1. Introduction

Transient thermal processes for silicon doping and solar cells fabrication by using pulsed laser irradiation or halogen lamps from gas or solid sources have been extensively investigated in these last years [1–4]. Indeed, pulsed laser annealing (PLA) of silicon substrate coated by a film containing dopants can be used to melt locally a near surface region and to produce significant incorporation of dopants from the surface into the molten surface layer.

The dopant distribution is driven by a simple liquid state diffusion [1, 5]. This feature gives an advantage in forming highly-doped shallow junctions but cannot be used for large scale silicon solar cells because of the limited treated area (localized doping).

On the other hand, it is well established now that rapid thermal diffusion (RTD) of dopants from doped spin-on glass films (SOD) using halogen lamps as heating source allows the formation of well controlled homogeneous junctions. The surface concentration and junction depth are easily controlled by doping source, processing temperature and time [6, 7]. This process has been successfully applied to the formation of low-cost, low-thermal budget silicon solar cells. Efficiencies as high as 17–18% [8] have been obtained for processing times less than 3 min.

To reach higher efficiencies and to allow cheaper metal contacting on silicon solar cells, the so called selective emitter structure, made of localized highly doped emitters for metal contacting and of shallow junctions to increase light conversion, together with back-surface fields regions are needed. Our challenge here is to combine both the above mentioned techniques (PLA + RTD) to keep all the features necessary for high efficiency and in particular lightly doped passivated emitter (SOD + RTD) and highly doped emitters beneath the metallic contacts (SOD + PLA). The two-step process described in this paper could of

---

* Corresponding author. Tel.: + 39-6-3048-6640; fax: + 39-6-3048-6405; e-mail: besi@casaccia.enea.it

course be extended beyond photovoltaics, to other large area electronic processes.

2. Experimental

Most of the experiments have been performed on CZ or on high quality FZ silicon, boron doped 1 Ω cm, ⟨100⟩ orientation, 500 to 525 μm thick, one side mirror like.

The spin-on glass (SOG) source was the solution P509 produced by Filmtronics (USA), with a concentration of 2×10^{21} P/cm^3, diluted with methanol. The thickness of the oxide films has been usually kept around 30 nm by adjusting spinning parameters.

After solvent evaporation (200°C for 10 min), a doped layer was formed in a rapid thermal furnace FAV4 JIPELEC (France), with the temperature in the range 800 to 900°C and time ranging from 30 to 60 s.

Some tests on laser doping have been carried out on untreated (only solvent evaporation, $T = 200$°C) or partially treated wafers ($T \ll 800$°C).

The sheet resistance of the emitters obtained by RTPs was in the range 100–150 Ω/□, a good interval for cell processing. In addition, an RTP co-diffusion of P-doped SOG on the front and of aluminum on the back of some cells was tried.

Laser overdoping was achieved by using a Q-switched Nd-YAG laser, frequency doubled (532 nm); this wavelength allows a good near-surface processing, as the silicon absorption coefficient is 8300 cm^{-1}, with a penetration depth of about 2 μm; at this wavelength, the minimum fluence value for melting is 2.0 J/cm^2.

Samples were positioned on motorized X–Y slits. Most of the experiments have been made by using an AUREL Nd-Yag laser, with an output power ranging from 0.5 to 0.9 W, repetition frequency from 1 to 5 kHz, pulse duration of about 90 ns and a maximum translation speed of 20 mm/s; a second Nd system from PHASE was also used, with slightly different setting values. The laser spot was kept around 30–40 μm in diameter, so as to have a good overlapping (of the order of 75%) and an easy grid process; large patterns made for measurements were 1 × 1 cm squares, obtained by multiple passages (lateral overlapping of 20%).

Our choice for a solid state Nd laser is due to the fact that this kind of apparatus gives overdoped emitters that can be easily 2–3 μm deep, a value that fits the requirements of screen printing contacting, perhaps the only contacting technique employed in photovoltaic large scale production.

Microelectronic equipment and high vacuum facilities have been generally used in our study to process samples for solar cells fabrication. This approach has been intentionally adopted in order to test the full po-

tential of our combined method, without introducing unwanted variables due to other processes with lower quality; the fabrication of cells by using screen printing for the contact formation has been successfully set up.

3. Results and discussion

The doping induced by a laser is a very fast process, of the order of 1 μs, that involves first of all the melting of the treated region, with a depth that is a function of laser wavelength and fluence. The dopant species are supplied by the pyrolysis of the atoms present on the surface. Due to the very high diffusivity of atoms in the liquid phase ($D = 10^{-4}$ cm^2/s), dopants are able to travel almost the whole length of the melted area prior to silicon solidification.

In Fig. 1 we show the sheet resistance versus laser fluence, E_{max}/e. In these samples, we have used a SOD film RTP processed at different annealing temperatures; from Fig. 1 it is possible to see that we can vary the sheet resistance value over a wide range of values, down to almost 10 Ω/□. Always in Fig. 1, the arrow on the y-axis shows the lower sheet resistance value obtained by using RTP alone, so that the large increase in the conductivity, induced by the laser, is clear.

Besides impinging laser power, the amount of dopants that can be moved into silicon is influenced by several parameters. Due to the fact that doping can be seen as a sequence of single shots, the laser repetition rate and translation slit speed determine the number of shots per point and thus the number of atoms pushed into the semiconductor. On the other side, film thickness, dilution and the treatment of the SOD film give

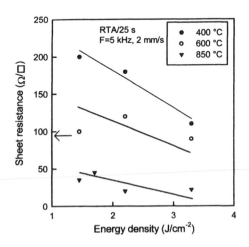

Fig. 1. Plot of sheet resistance versus fluence E_{max}/e for several film treatments. Each SOD film was RTP processed for 25 s. Marked by arrow is the lower sheet resistance before laser processing.

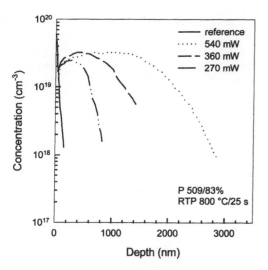

Fig. 2. Phosphorus SIMS profiles for several laser powers. Reference RTP (laser untreated) sample is also shown (solid line).

the number of doping atoms available from the source: we found, for example, that undensified films (only 200°C for 10 min to get rid of solvents) tend to evaporate under laser beam, while densified (RTP or classical temperature process) SOD is more stable and leads to a large incorporation of phosphorus atoms into silicon. The right combination of the above mentioned variables can lead to the formation of a box-like junction. In Fig. 2, we show the doping profiles of overdoped regions at three laser powers, also shown is the profile of the untreated reference sample.

Several fundamental questions about the re-crystallized material can arise when a very short timescale is involved, as in the case of laser doping. In fact, crystalline disorientation and large defect formation could be induced by the process, lowering the quality of the junction and of the devices.

In Fig. 3 the diffraction pattern obtained, respectively, from the silicon substrate and the laser treated area is shown. It is quite easy to see that the solidification of the molten phase is epitaxial, with the same

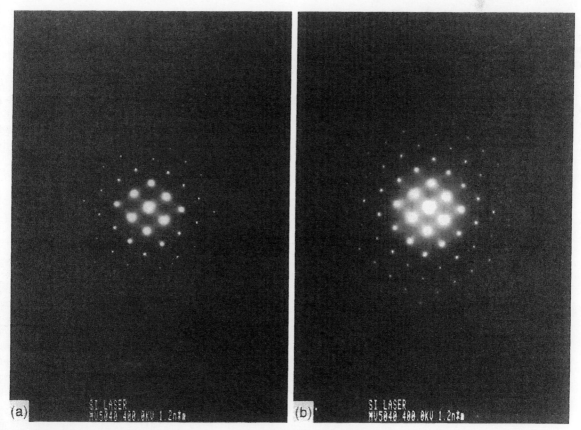

Fig. 3. Diffraction patterns obtained from bulk crystalline silicon (left) and from re-crystallized material (right). Bulk silicon was FZ grown, ⟨100⟩ oriented.

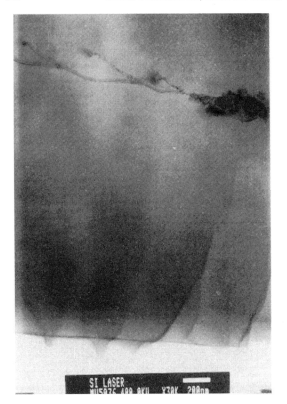

Fig. 4. Cross-section TEM picture. From bottom: vacuum–silicon interface, re-crystallized region (almost 1.5 μm thick), dislocations at melted-bulk interface.

orientation as the underlying material. Fig. 4 shows the cross-section TEM picture of the same sample: the molten region at about 1.5 μm depth is well recognized by means of the presence of a dislocated area at the original liquid–solid interface.

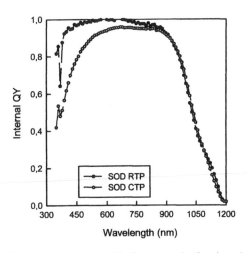

Fig. 5. Internal quantum yield of two sample. One is a classically (CTP, 900°C/30 min) treated cell, the other one is RTP (850°C/25 s) processed.

SEM and SEM EBIC analysis give us information about the morphology and electrical activity of the laser formed junctions, usually showing (EBIC) the overdoped region as an area with a slightly higher carrier recombination. This fact can be explained by the different solubility of phosphorus in the liquid and in the solid phase and by the segregation of the doping element during silicon solidification, starts from the liquid–solid interface and moves toward the surface [9].

Turning now to silicon solar cells, conversion efficiency of our devices lies around 16% (16.3% maximum value). Main photovoltaic parameters, as taken from I–V light curves, say that the open circuit voltage is in the range 600–610 mV, the fill factor usually varying from 74 to 78, while the selective emitter structure of the device allows high values of short circuit current.

Dark I–V curves of our devices do not evidence detrimental effects due to laser processing and data analysis show very low series resistance values, of the order of 50–70 mΩ.

In Fig. 5 we show the curves of the internal spectral response of two samples, one with an emitter of about 100 Ω/□ made by using a SOD diffused in classical furnace, the other with an emitter of 130–150 Ω/□ prepared by RTP; as it is possible to see, there is a better conversion of the low wavelength radiation for the device processed by RTP, undoubtedly due to the shallower emitter and only the infrared portion of both curves gives the same signal of quantum yields, as it is a function of the bulk.

The applicability of our process to screen printing steps has also been tested [10] with success. Laser overdoped regions meet the requirements of screen printing technique to have deep and heavily doped regions to create contacts, so that it is possible to apply this widely used contacting process to very shallow doped areas.

4. Conclusions

The well known capability of rapid thermal processing to obtain high quality shallow emitters can be successfully combined with laser overdoping, thus having a two step process that is able create high/low junctions.

Laser-induced doping allows precise control of the amount of atoms incorporated and, at the same time, it does not jeopardize the quality of the junction through massive creation of defects.

The combined process that we have just described is particularly suitable to making contact patterns on large area devices.

In our study, we have applied RTP plus laser to solar cell fabrication, in order to get an effective selective emitter structure and high efficiency devices. Results obtained so far show that this process is able to increase photovoltaic performances of silicon solar cells without using sophisticated microelectronic processes.

Acknowledgements

We would like to acknowledge Marco Vittori of ENEA for his TEM analysis. Work supported by E.C. in the Joule III 'Lowthermcells' Program, contract No. JOUR3-CT95-0069.

References

[1] Bentini GG, Bianconi M, Correra L, Nipoti R, Summonte C, Cohen C, Siejka J. E-MRS Meeting Conf. Proc., vol. XV. 1987. p. 251.

[2] Bentini GG, Bianconi M, Summonte C. Appl Phys A 1988;45:317.

[3] Slaoui A, Elliq M, Pattyn H, Fogarassy E, de Unamuno S, Stuck R. MRS Conf Proc 1991;236:389.

[4] Slaoui A, Foulon F, Siffert P. J Appl Phys 1990;67:6197.

[5] Elliq M, Slaoui A, Fogarassy E, Pattyn H, Siffert P. MRS Conf Proc 1991;219:739.

[6] Lachiq A, Slaoui A, Georgopoulos L, Ventura L, Monna R, Muller JC. Progress in photovoltaics: research and applications, Vol. 4. 1996. p. 329.

[7] Doshi P, Mejia J, Tate K, Kamra S, Rohatgi A, Narayan S, Singh R. 25th IEEE PVSC Conf. Proc. 1996. p. 421.

[8] Rohatgi A, Doshi P, Kamra S. 14th European PVSECE Conf. Proc. 1997. p. 660.

[9] Pirozzi L, Besi-Vetrella U, Salza E. Laser Applications in Microelectronic and Optoelectronic Manufacturing II SPIE Conf. Proc. 1997. p. 119.

[10] Besi-Vetrella U, Pirozzi L, Salza E, Ginocchietti G, Ferrazza F, Ventura L, Slaoui A, Muller JC. 26th IEEE PVSC Conf. Proc. 1997. p. 135.

PERGAMON

Materials Science in Semiconductor Processing 1 (1998) 331–334

MATERIALS
SCIENCE IN
SEMICONDUCTOR
PROCESSING

Rapid thermal annealing applied to the optimization of titanium oxide arc

M. Lemiti *, J.P. Boyeaux, H. El Omari, A. Kaminski, A. Laugier

Laboratoire de Physique de la Matière, UMR CNRS 5511, Institut National des Sciences Appliquées de Lyon, 20 Avenue Albert Einstein, 69621 Villeurbanne Cedex, France

Abstract

In this work, rapid thermal annealing (RTA) in the 400–700°C range is used to adjust simultaneously the index and thickness of titanium oxide layers in order to achieve an antireflection coating (ARC) in industrial conditions (large solar cells with a high throughput capability). The technique used for high production rate and low cost coating process is the atmospheric pressure chemical vapor deposition (APCVD) at low temperature.

Titanium oxide layers are obtained from the hydrolysis of two precursors, namely the tetraisopropoxide titanate (TPT) and the titanium tetrachloride. In the first case, on silicon substrates at 150°C, we have obtained refractive indices of 1.8–1.9 with layer thicknesses in the range 70–100 nm. In the second case, at 100°C, the refractive index is 2.1–2.2 as deposited and the thickness in the same range. After RTA lasting 120 s at 700°C, the refractive index is 2.25 and the thickness is decreased by 40% with the first precursor, while with chloride, the index increases until 2.4 and the thickness is decreased by 30%. These results are discussed by means of X-ray diffraction and SIMS analysis. © 1999 Elsevier Science Ltd. All rights reserved.

1. Introduction

Rapid thermal processing (RTP) using incoherent light is very interesting for the reduction of the thermal budget, gas consumption and fabrication time of devices. Particularly, this technique has been successfully used for the fabrication of solar cells on monocrystalline [1], as well as on polycrystalline, silicon [2]. In the same way, the capability to sinter the screen printed contacts by rapid thermal annealing (RTA) instead of conventional sintering has been previously shown [3]. RTA can be used without degrading the transport properties of the solar cell base region, particularly the minority carrier diffusion length [4, 5].

On the other hand, the reflectance of the Si surface can be reduced to less than 3% for most of the concerned spectrum range when the outside medium is air: for example, an optimized MgF_2/ZnS double layer antireflection coating (ARC) has been used [6].

However, the realization of layers such as MgF_2 and ZnS by vacuum deposition cannot be performed in industrial conditions for large solar cells (10×10 cm^2) with a high throughput (one cell per second). The technique used for a high volume and low cost coating process is the atmospheric pressure chemical vapor deposition (APCVD) at low temperature and films such as titanium dioxide are currently deposited [7, 8].

The aim of this study is the development of an antireflective coating technique with good optical properties associated to a high rate of fabrication. In this way, indices and thicknesses of ARC are optimized by RTA.

2. Experimental

2.1. Atmospheric pressure chemical vapor deposition of oxide layers

In the case of APCVD, two methods can be used from organometallic vapor, namely the pyrolysis or

---

* Corresponding author.

the hydrolysis of the titanium isopropoxide Ti(OC$_3$H$_7$)$_4$ (TPT). In the first one, vapors delivered by nitrogen as carrier gas are dispersed uniformly across the belt by the injector head and the spray of TPT deposited on cells gives the titanium dioxide film by thermal decomposition. In the second method, a water vapor delivery system is required and it is necessary to have 2 moles of water vapor for 1 mole of TPT in order to obtain a complete reaction.

It is also possible to hydrolyse another precursor like titanium tetrachloride TiCl$_4$ to prepare TiO$_2$ films and we have investigated this method in order to achieve a higher optical index associated to a better deposition rate.

2.2. Rapid thermal annealing

The RTA treatments were carried out in a 30 kW incoherent light annealing furnace. The system consists of a quartz chamber flowed with gas like Ar, N$_2$ or air, surrounded by water-cooled reflecting walls and heated by 12 tungsten halogen tubes. The thermal cycles are controlled by programming the power applied to the lamps and a data acquisition system allows a real-time recording of the sample temperature via a type K thermocouple cemented in a reference sample. Typical RTA cycles begin with a fast heating stage of about 100°C/s, followed by a temperature plateau (up to 1200°C) of a few seconds. Cooling rates in the 5 to 100°C/s range can be achieved.

2.3. Characterization of oxide layers

The index of refraction and the thickness measurements are made with a routine ellipsometer working with a fixed incidence angle and a single wavelength, the standard source being a HeNe laser at 632.8 nm wavelength; structural analysis correlated with index measurements are performed by X-ray diffraction and the quality of the deposit is carried out by SIMS profiles.

3. Results

3.1. Deposition of layers

Optimization of the indices and thicknesses of ARC is performed by theoretical simulation. The thin layer of passivation (SiO$_2$ with 15 nm thickness and 1.45 as index) and the encapsulated layer with 1.45 as a refractive index are taken into account. Then, the value of the index to approach is 2.4 with a thickness calculated at half–quarter wavelength of about 60 nm. However, a problem in multicrystalline solar cells is the texturization of the surface avoiding simultaneous optimiz-

ation on each grain of the device. In order to avoid any difficulty in optimization of layers, the substrates used in this work are monocrystalline CZ silicon, p-type, 2–5 Ω cm and (111) oriented.

For titanium dioxide films deposited above 150°C from hydrolysis of TPT on heated silicon substrates, we have obtained refractive indices of 1.8–1.9 as deposited, with a layer thickness of 100 nm in order to anticipate the thickness decrease which occurs during thermal treatments necessary for solar cell fabrication.

On the other hand, using the hydrolysis of TiCl$_4$, thin TiO$_2$ films are produced at a temperature as low as 80°C. The refractive index is 2.1–2.2 as deposited and the thickness in the range 70–90 nm. A relatively low temperature of substrates is used in order to keep some porosity of the layer, because front contact realisation is performed after ARC deposition by the firing through process.

3.2. Optimization of layers parameters by RTA

The basic idea is to use thermal treatments necessary to solar cell fabrication, like contact realization [3, 9], in order to optimize the index and thickness of ARC. Rapid thermal annealing-up to 700°C during 120 s—is performed to simulate the industrial belt furnace and this thermal treatment is carried out in air in order to facilitate the industrial use. The main advantage of RTA is that temperature profiles are well defined with accuracy compared to the typical cycle obtained with classical thermal treatment. In this case, the heating and cooling step duration is too long compared to the 120 s duration of the temperature plateau. We use 10°C/s as a cooling down rate because this seems to be the best rate for both front and back contacts [10]. For TiO$_2$ films, annealing in air causes crystallization: anatase is formed at 350°C and rutile at 750°C. Bulk TiO$_2$ indices are 2.56 for anatase and 2.75 for rutile [11]. During heat treatments, the density and then the refractive index increase substantially with increasing annealing temperature and the film thickness decreases also.

Fig. 1 gives variations of the refractive index n versus annealing temperature of titanium dioxide layers deposited on silicon wafers submitted to RTA lasting 120 s. Three typical examples are shown: two TiO$_2$ layers from TPT, with values of 1.88 and 1.9 as initial indices (substrate temperature 150°C) and a TiO$_2$ film from TiCl$_4$ with 2.14 as an initial index (substrate temperature 100°C). Fig. 2 gives variations of dioxide layer thickness versus annealing temperature submitted to RTA during 120 s under air. Two layers are considered, one from TPT and one from chloride, with the same conditions as Fig. 1. After annealing, from TPT, the refractive index variation is 20% and the thickness is decreased by 40%. With chloride, the index increases

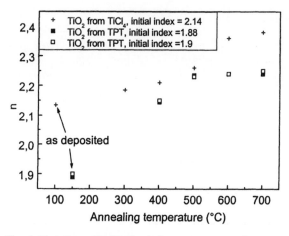

Fig. 1. Variations of refractive index n versus annealing temperatures of layers submitted to RTA under air during 120 s.

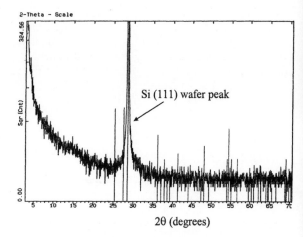

Fig. 3. X-ray diffraction spectrum ($2\theta/\theta$ mode) of TiO$_2$ layer upon silicon substrate (111) before annealing.

until 2.38 and the thickness decreases by 30%: a refractive index up to 2.55 at 540 nm has been obtained with this precursor [12].

4. Discussion and concluding remarks

These results can be tentatively analysed by X-ray diffraction and SIMS experiments. All films are amorphous as grown, but small crystals of anatase and rutile can also exist. These crystallites are more frequent if the thickness of the thin layer is higher and/or if the substrate temperature is raised [13]. Fig. 3 shows the X-ray diffraction spectrum in $2\theta/\theta$ mode of an oxide layer obtained with chloride on (111) silicon substrate before annealing. Fig. 4 shows the same spectrum after annealing and the inset shows a beginning

of the anatase phase ($2\theta = 25.28°$) explaining the increase of the index.

On the other hand, the same annealing (700°C, 120 s) gives an index of about 2.4 with TiCl$_4$ precursor and, with TPT, the index seems limited to 2.25. As pointed out in the literature (see, for example, Ref. [14]), acid or base catalysis can be used to promote decoupling between hydrolysis and condensation: hydrolysis rates increase under acid conditions, while condensation proceeds faster in the presence of base catalysts. Titanium isopropoxide is rapidly and completely hydrolyzed under acid conditions before condensation begins. Fully hydroxylated precursors then lead to crystalline (anatase or rutile) TiO$_2$. Under neutral or basic conditions, hydrolysis and condensation occur simultaneously giving molecular species. Some ligands are still present in the nuclei preventing the

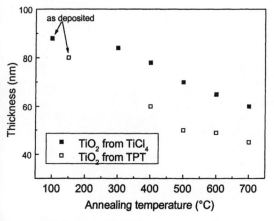

Fig. 2. Variations of thickness of the oxide layers versus annealing temperatures of layers submitted to RTA under air during 120 s.

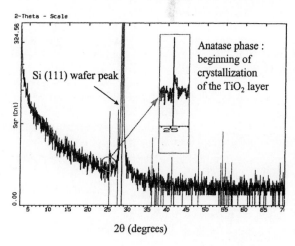

Fig. 4. Same spectrum as Fig. 3 after annealing at 700°C, 120 s under air. The inset shows a beginning of the anatase phase ($2\theta = 25.28°$).

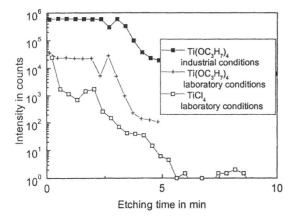

Fig. 5. SIMS profiles of Na$^+$ ion in three TiO$_2$ layers: two layers from TPT (industrial and laboratory conditions) and one from titanium chloride.

Acknowledgements

This work was supported by the CNRS-ECODEV, ADEME and by CEC Joule III contract (Menhir JOR3-CT95-0094) with the collaboration of Photowatt International S.A., 38300, Bourgoin-Jallieu, France. Authors would like to thank Dr. C. Dubois for SIMS experiments.

crystallization of pure TiO$_2$. Aggregated amorphous powders are obtained, the morphology of which depends on the pH of the solution. Then, acid and base species are always searched in oxide layers when SIMS analyses are performed.

Fig. 5 shows a comparison of three SIMS profiles relative to Na$^+$ ion for three types of TiO$_2$ layers: a layer obtained in industrial conditions from TPT; a film deposited in laboratory conditions from TPT and a coating performed by TiCl$_4$ hydrolysis. The sodium content is more important in TiO$_2$ layers realized from TPT, with flat profiles of Na$^+$. We think that the presence of ionic species like Na$^+$ in the dioxide layer prevents its crystallization during the annealing treatment and then limits the increase of the index.

The origin of Na is sought for both the cases of industrial and laboratory conditions. In the last case, a clean room is not used for layer deposition and sodium is a well known contaminant. The concentration of sodium in industrial TPT is initially 3 ppm and decreases to 2.3 ppm after 30 layer fabrications. In the industrial deposition, the effects of NaOH texturization and the contamination by alkaline species during thermal treatments (POCl$_3$ diffusion and SiO$_2$ passivation layer) up to 850°C in a fused silica tube can be suspected.

However, we can obtain layers with adjustable index suitable for an industrial realization of ARC.

References

[1] Schindler R, Reis I, Wagner B, Eyer A, Lautenschlager H, Schetter C, Warta W, Hartiti B, Slaoui A, Muller JC, Siffert P. 23rd IEEE Photovoltaic Specialist Conf. Record. 1993. p. 162.

[2] Hartiti B, Slaoui A, Muller JC, Siffert P, Schindler R, Reis I, Wagner B, Eyer A. 23rd IEEE Photovoltaic Specialist Conf. Record. 1993. p. 224.

[3] El Omari H, Boyeaux JP, Laugier A. 1st World Conf. on Photovoltaic Energy Conversion. 24th IEEE Photovoltaic Specialist Conf. Record. 1994. p. 1539.

[4] Quat VT, Eichhammer W, Siffert P. Appl Phys Lett 1988;53:1928.

[5] Boyeaux JP, Masri K, Chaussemy G, Mayet L, Laugier A. 21st IEEE Photovoltaic Specialist Conf. Record. 1990. p. 629.

[6] Zhao J, Green MA. IEEE Trans Electron Devices 1991;38:1925.

[7] Wong DC, Waugh A, Yui B, Sharrock P. 1st World Conf. on Photovoltaic Energy Conversion. 24th IEEE Photovoltaic Specialist Conf. Record. 1994. p. 905.

[8] Wong DC, Waugh A. Mater Res Soc Symp Proc 1996;426:503.

[9] Laugier A, El Omari H, Boyeaux JP, Hartiti B, Muller JC, Nam Le Quan, Sarti D. 1st World Conf. on Photovoltaic Energy Conversion. 24th IEEE Photovoltaic Specialist Conf. Record. 1994. p. 1535.

[10] El Omari H, Boyeaux JP, Laugier A. 25th IEEE Photovoltaic Specialist Conf. Record. 1996. p. 585.

[11] Yokozawa M, Iwasa H, Teramoto I. Jpn J Appl Phys 1968;7:96.

[12] Feuersanger AE. Proc IEEE 1964;52:1463.

[13] Lottiaux M, Boulesteix C, Nihoul G, Varnier F, Flory F, Galindo R, Pelletier E. Thin Solids Films 1989;170:107.

[14] Livage J, Henry M, Sanchez C. Prog Solid State Chem 1988;18:259.

MATERIALS SCIENCE IN SEMICONDUCTOR PROCESSING

Contents Select
A focused survey of recently published literature

Papers published in the field of semiconductor processing are scattered throughout a wide range of journals and keeping track of all that is published is not easy. In recognition of this the Editors and members of the Editorial Board of Materials Science in Semiconductor Processing *have scanned the recently published literature to identify papers that are closely related to the scope of the journal and of potential interest to you as a reader. The list is of course selected and cannot claim to be comprehensive. Nonetheless, we hope that it will help to bring to your attention papers that otherwise you may have missed.*

If you have recently published a paper related to the scope of the journal but do not see your paper listed in this section, please send an e-mail to p.mestecky@elsevier.co.uk with the full reference. All submissions will be then considered for inclusion in the next issue.

APPLIED PHYSICS LETTERS

Koichi Takemura, Takehiro Noguchi, Takashi Hase, Yoichi Miyasaka, Dielectric anomaly in strontium bismuth tantalate thin films, *Appl. Phys. Letts*, vol **73** (12) p. 1649 (1998).

D. A. Buchanan, F. R. McFeely, J. J. Yurkas, Fabrication of midgap metal gates compatible with ultrathin dielectrics *Appl. Phys. Letts*, vol **73** (12) p. 1676 (1998).

H. Rücker, B. Heinemann, W. Röpke, R. Kurps, D. Kürger, G. Lippert, H. J. Osten, Suppressed diffusion of boron and carbon in carbon-rich silicon, *Appl. Phys. Letts*, vol **73** (12) p. 1682 (1998).

A. Ural, P. B. Griffin, J. D. Plummer, Experimental evidence for a dual vacancyinterstitial mechanism of self-diffusion in silicon, *Appl. Phys. Letts*, vol **73** (12) p. 1706 (1998).

Xianfang Zhu, J. S. Williams, J. C. McCallum, Surface morphological structures in ultra-high-dose self-implanted silicon, *Appl. Phys. Letts*, vol **73** (13) p. 1811 (1998).

Jae Hyuk Jang, Ki Hyun Yoon, Hyun Jung Shin, Electric fatigue in solgel prepared $Pb(Zr,Sn,Ti)NbO_3$ thin films, *Appl. Phys. Letts*, vol **73** (13) p. 1823 (1998).

M. Copel, J. D. Baniecki, P. R. Duncombe, D. Kotecki, R. Laibowitz, D. A. Neumayer, T. M. Shaw, Compensation doping of $Ba_{0.7}Sr_{0.3}TiO_3$ thin films, *Appl. Phys. Letts*, vol **73** (13) p. 1832 (1998).

Katsuyoshi Endo, Kenta Arima, Toshihiko Kataoka, Yasushi Oshikane, Haruyuki Inoue, Yuzo Mori, Atomic structures of hydrogen-terminated Si(001) surfaces after wet cleaning by scanning tunneling microscopy, *Appl. Phys. Letts*, vol **73** (13) p. 1853 (1998).

Nobuyuki Ikarashi, Analytical transmission electron microscopy of hydrogen-induced degradation in ferroelectric $Pb(Zr, Ti)O_3$ on a Pt electrode, *Appl. Phys. Letts,* vol **73** (14) p. 1955 (1998).

Yongfei Zhu, Jinsong Zhu, Yoon J. Song, S. B. Desu, Laser-assisted low temperature processing of $Pb(Zr, Ti)O_3$ thin film, *Appl. Phys. Letts*, vol **73** (14) p. 1958 (1998).

F. Bònoli, M. Iannuzzi, Leo Miglio, V. Meregalli, Electronic origin of the stability trend in $TiSi_2$ phases with Al or Mo layers, *Appl. Phys. Letts,* vol **73** (14) p. 1964 (1998).

SIMOX MOSFET's, *IEEE Trans. Elect. Dev.*, vol **45** (10) p. 2146 (1998).

J.-H. Lee, J.-S. Lyu, T. M. Roh, B. W. Kim, Effects of buffer layer structure on polysilicon buffer LOCOS for the isolation of submicron silicon devices, *IEEE Trans. Elect. Dev.*, vol **45** (10) p. 2153 (1998).

IEEE TRANSACTIONS ON SEMICONDUCTOR MANUFACTURING

M. Zhou, M. D. Jeng, Modeling, analysis, simulation, scheduling, and control of semiconductor manufacturing systems: a Petri Net approach, *IEEE Trans. Semicon. Manuf.*, vol. **11** (3) p. 333 (1998).

M. D. Jeng, X. Xie, S. W. Chou, Modeling, qualitative analysis, and performance evaluation of the etching area in an IC wafer fabrication system using Petri Nets, *IEEE Trans.Semicon. Manuf.*, vol. **11** (3) p. 358 (1998).

M. Allam, H. Alla, Modeling and simulation of an electronic component manufacturing system using hybrid Petri Nets, *IEEE Trans. Semicon. Manuf.*, vol. **11** (3) p. 374 (1998).

J. D. Stuber, I. Trachtenberg, T. F. Edgar, Design and modeling of rapid thermal processing systems, *IEEE Trans.Semicon. Manuf.*, vol. **11** (3) p. 442 (1998).

S. R. Runnels, I. Kim, J. Schleuter, C. Karlsrud, M. Desai, A modeling tool for chemical-mechanical polishing design and evaluation, *IEEE Trans.Semicon. Manuf.*, vol. **11** (3) p. 501 (1998).

R. N. Wall, M. C. Olewine, R. Augur, J. DiGregorio, G. Colovos, A new four-level metal interconnect system tailored to an advanced 0.5-lm BiCMOS technology, *IEEE Trans. Semicon. manuf.*, vol. **11** (4) p. 624 (1998).

J. C. Chiou, J. Y. Yang, A CVD epitaxial deposition in a vertical barrel reactor: process modeling using cluster-based fuzzy logic models, *IEEE Trans.Semicon. Manuf.*, vol. **11** (4) p. 645 (1998).

JOURNAL OF APPLIED PHYSICS

M. E. Law, Y. M. Haddara, K. S. Jones, Effect of the silicon/oxide interface on interstitials: Di-interstitial recombination, *J. Appl, Phys.*, vol **84** (7) p. 3555 (1998).

W. Götz, G. Pensl, W. Zulehner, R. C. Newman, S. A. McQuaid, Thermal donor formation and annihilation at temperatures above $500\,^{\circ}C$ in Czochralski-grown Si, *J. Appl, Phys.*, vol **84** (7) p. 3561 (1998)

P. M. Rousseau, P. B. Griffin, W. T. Fang, J. D. Plummer, Arsenic deactivation enhanced diffusion: A time, temperature, and concentration study, *J. Appl, Phys.*, vol **84** (7) p. 3593 (1998).

Hanchen Huang, George H. Gilmer, Tomas Díaz de la Rubia, An atomistic simulator for thin film deposition in three dimensions, *J. Appl, Phys.*, vol **84** (7) p. 3636 (1998).

Shyam Ramalingam, Dimitrios Maroudas, Eray S. Aydil, Interactions of SiH radicals with silicon surfaces: An atomic-scale simulation study, *J. Appl, Phys.*, vol **84** (7) p. 3895 (1998).

P. I. Gaiduk, V. S. Tishkov, S. Yu. Shiryaev, A. Nylandsted Larsen, Effect of composition and annealing on structural defects in high-dose arsenic-implanted $Si_{1-x}Ge_x$ alloys, *J. Appl, Phys.*, vol **84** (8) p. 4185 (1998).

N. Otsuka, M. Kito, M. Ishino, Y. Matsui, F. Toujou, Control of double diffusion front unintentionally penetrated from a Zn doped InP layer during metalorganic vapor phase epitaxy, *J. Appl, Phys.*, vol **84** (8) p. 4239 (1998).

M. Koizuka, H. Yamada-Kaneta, Gap states caused by oxygen precipitation in Czochralski silicon crystals, *J. Appl, Phys.*, vol **84** (8) p. 4255 (1998).

B. I. Boyanov, P. T. Goeller, D. E. Sayers, R. J. Nemanich, Film thickness effects in the $CoSi_{1-x}Ge_x$ solid phase reaction, *J. Appl, Phys.*, vol **84** (8) p. 4285 (1998).

J. L. Benton, K. Halliburton, S. Libertino, D. J. Eaglesham, S. Coffa, Electrical signatures and thermal stability of interstitial clusters in ion implanted Si, *J. Appl, Phys.*, vol **84** (9) p. 4749 (1998).

Masashi Uematsu, Simulation of clustering and transient enhanced diffusion of boron in silicon, *J. Appl, Phys.*, vol **84** (9) p. 4781 (1998).

S. W. King, J. P. Barnak, M. D. Bremser, K. M. Tracy, C. Ronning, R. F. Davis, R. J. Nemanich, Cleaning of AlN and GaN surfaces, *J. Appl, Phys.*, vol **84** (9) p. 5248 (1998).

Y.-L. Shen, Stresses, deformation, and void nucleation in locally debonded metal interconnects, *J. Appl, Phys.*, vol **84** (10) p. 5525 (1998).

V. Krishnamoorthy, K. Moller, K. S. Jones, D. Venables, J. Jackson, L. Rubin, Transient enhanced diffusion and defect microstructure in high dose, low energy As + implanted Si, *J. Appl, Phys.*, vol **84** (11) p. 5997 (1998).

N. Vasudevan, R. B. Fair, H. Z. Massoud, T. Zhao, K. Look, Y. Karpovich, M. J. Hart, ON-state reliability of amorphous-silicon antifuses, *J. Appl, Phys.,* vol **84** (11) p. 6440 (1998).

Yong Sun, Tatsuro Miyasato, Nobuo Sonoda, Outdiffusion of the excess carbon in SiC films into Si substrate during film growth, *J. Appl, Phys.,* vol **84** (11) p. 6451 (1998).

JOURNAL OF ELECTRONIC MATERIALS

Q. Chen, M. Hillert, B. Sundman, W. A. Oates, S. G. Fries, R. Schmid-Fetzer, Phase equilibria, defect chemistry and semiconducting properties of CdTe(s)–thermodynamic modeling, *J. Elec. Mater.,* vol. **27** (8) p. 961 (1998).

K. B. Jung, J. Hong, H. Cho, J. R. Childress, S. J. Pearton, M. Jenson, A. T. Hurst, Jr., Plasma chemistries for dry etching of NiFe and NiFeCo, *J. Elec. Mater.,* vol. **27** (8) p. 972 (1998).

Haksoo Han, Hyunsoo Chung, Yung-Il Joe, Seongsu Park, Gwangchong Joo, Nam Hwang, Minkyu Song, The application of flip-chip bonding interconnection technique on the module assembly of 10 Gbps laser diode, *J. Elec. Mater.,* vol. **27** (8) p. 985 (1998).

W. A. Doolittle, S. Kang, T. J. Kropewnicki, S. Stock, P. A. Kohl, A. S. Brown, MBE growth of high quality GaN on LiGaO$_2$, *J. Elec. Mater.,* vol. **27** (8) p. L58 (1998).

Xiang Lu, S. Sundar Kumar Iyer, Jin Lee, Brian Doyle, Zhineng Fan, Paul K. Chu, Chenming Hu, Nathan W. Cheung, Plasma immersion ion implantation for SOI synthesis: SIMOX and ion-cut, *J. Elec. Mater.,* vol. **27** (9) p. 1059 (1998).

John G. Darab, Dean W. Matson, Continuous hydrothermal processing of nano-crystalline particulates for chemical-mechanical planarization, *J. Elec. Mater.,* vol. **27** (10) p. 1068 (1998).

Thomas Bibby, Karey Holland, Endpoint detection for CMP, *J. Elec. Mater.,* vol. **27** (10) *J. Elec. Mater.,* vol. **27** (10) p. 1073 (1998).

C. Rogers, J. Coppeta, L. Racz, A. Philipossian, F.B. Kaufman, D. Bramono, Analysis of flow between a wafer and pad during CMP processes, *J. Elec. Mater.,* vol. **27** (10) p. 1082 (1998).

Dan Towery, Michael A. Fury, Chemical mechanical polishing of polymer films, *J. Elec. Mater.,* vol. **27** (10) p. 1088 (1998).

Ahmed A. Busnaina, T.M. Elsawy, Post-CMP cleaning using acoustic streaming, *J. Elec. Mater.,* vol. **27** (10) p. 1095 (1998).

Jianshe Tang, David Dornfeld, Suzette Keefe Pangrle, Alvin Dangca, In-process detection of microscratching

during CMP using acoustic emission sensing technology, *J. Elec. Mater.,* vol. **27** (10) p. 1099 (1998).

D. Kumar, Ali Ata, Uday Mahajan, Rajiv K. Singh, Role of line-beam on the removal of particulate contaminations from solid surfaces by pulsed laser, *J. Elec. Mater.,* vol. **27** (10) p. 1104 (1998).

Krishna Rajan, Mechanical processes in chemical-mechanical planarization: plasticity effects in oxide thin films, *J. Elec. Mater.,* vol. **27** (10) p. 1107 (1998).

S. Soloviev, I. Khlebnikov, V. Madangarli, T. S. Sudarshan, Correlation between oxide breakdown and defects in SiC wafers, *J. Elec. Mater.,* vol. **27** (10) p. 1124 (1998).

V. Khemka, T. P. Chow, R. J. Gutmann, Effect of reactive ion etch-induced damage on the performance of 4H-SiC Schottky barrier diodes, *J. Elec. Mater.,* vol. **27** (10) p. 1128 (1998).

S. Jin, H. Mavoori, Processing and properties of CVD diamond for thermal management, *J. Elec. Mater.,* vol. **27** (11) p. 1148 (1998).

S. K. Kang, S. Purushothaman, Study of interfacial reactions between tin and copper by differential scanning calorimetry, *J. Elec. Mater.,* vol. **27** (11) 1199 (1998).

X. G. Zhang, P. Li, G. Zhao, D. W. Parent, F. C. Jain, J. E. Ayers, Removal of threading dislocations from patterned heteroepitaxial semiconductors by glide to sidewalls, *J. Elec. Mater.,* vol. **27** (11) p.1248 (1998).

JAPANESE JOURNAL OF APPLIED PHYSICS

H. R. Huff, D. W. McCormack, C. Au, T. Messina, R. Goodall, Current status of 200 mm and 300 mm silicon Wafers, *Jpn. J. Appl. Phys.,* vol. 37, 1210 (1998)

JOURNAL OF VACUUM SCIENCE AND TECHNOLOGY A

Gabriel I. Font, Iain D. Boyd, Jitendra Balakrishnan, Effects of wall recombination on the etch rate and plasma composition of an etch reactor, *J. Vac. Sci. Technol. A,* vol **16** (4) p. 2057 (1998).

K. Yasutake, A. Takeuchi, H. Kakiuchi, K. Yoshii, Molecular beam epitaxial growth of AlN single crystalline films on Si (111) using radio-frequency plasma assisted nitrogen radical source, *J. Vac. Sci. Technol. A,* vol **16** (4) p. 2140 (1998).

J. Hong, J. A. Caballero, E. S. Lambers, J. R. Childress, S. J. Pearton, Inductively coupled plasma etch processes for NiMnSb, *J. Vac. Sci. Technol. A,* vol **16** (4) p. 2153 (1998).

JOURNAL OF VACUUM SCIENCE AND TECHNOLOGY B

phase epitaxy grown ZnTe layers on (001) GaAs, *J. Vac. Sci. Technol B*, Vol. **16** (5) p. 2650 (1998).

Xian-an Cao, Hai-tian Hu, Xun-min Ding, Ze-liang Yuan, Yang Dong, Xi-ying Chen, Bing Lai, Xiao-yuan Hou, Passivation of the GaAs(100) surface with a vapor-deposited GaS film, *J. Vac. Sci. Technol B*, Vol. **16** (5) p. 2656 (1998).

M. Losurdo, P. Capezzuto, G. Bruno, P. R. Lefebvre, E. A. Irene, Study of the mechanisms of GaN film growth on GaAs surfaces by thermal and plasma nitridation, *J. Vac. Sci. Technol B*, Vol. **16** (5) p. 2665 (1998)

X. Hue, B. Boudart, Y. Crosnier, Gate recessing optimization of GaAs/Al$_{0.22}$Ga$_{0.78}$As heterojunction field effect transistor using citric acid/hydrogen peroxide/ammonium hydroxide for power applications, *J. Vac. Sci. Technol B*, Vol. **16** (5) p. 2675 (1998).

J. T. Jones, E. T. Croke, C. M. Garland, O. J. Marsh, T. C. McGill, Epitaxial silicon grown on CeO$_2$/Si(111) structure by molecular beam epitaxy, *J. Vac. Sci. Technol B*, Vol. **16** (5) p. 2686 (1998).

J. Hong, H. Cho, T. Maeda, C. R. Abernathy, S. J. Pearton, R. J. Shul, W. S. Hobson, New plasma chemistries for dry etching of InGaAlP alloys: BI$_3$ and Bbr$_3$, *J. Vac. Sci. Technol B*, Vol. **16** (5) p. 2690 (1998).

A. Mitra, C. D. Nordquist, T. N. Jackson, T. S. Mayer, Magnetron ion etching of through-wafer via holes for GaAs monolithic microwave integrated circuits using SiCl$_4$, *J. Vac. Sci. Technol B*, Vol. **16** (5) p. 2695 (1998).

A. C. Westerheim, R. D. Jones, P. J. Mager, J. H. Dubash, T. J. Dalton, M. W. Goss, S. K. Baum, S. K. Dass, High-density, inductively coupled plasma etch of sub half-micron critical layers: Transistor polysilicon gate definition and contact formation, *J. Vac. Sci. Technol B*, Vol. **16** (5) p. 2699 (1998).

A. C. Westerheim, J. M. Bulger, C. S. Whelan, T. S. Sriram, L. J. Elliott, J. J. Maziarz, Integration of chemical vapor deposition titanium nitride for 0.25 lm contacts and vias, *J. Vac. Sci. Technol B*, Vol. **16** (5) p. 2729 (1998).

D. B. Knorr, S. M. Merchant, M. A. Biberger, Development of texture in interconnect thin film stacks, *J. Vac. Sci. Technol B*, Vol. **16** (5) p. 2734 (1998).

Yasushi Igarashi, Toshio Ito, Electromigration properties of copper-zirconium alloy interconnects, *J. Vac. Sci. Technol B*, Vol. **16** (5) p. 2745 (1998).

Tomoyuki Yoshida, Shoji Hashimoto, Yasuichi Mitsushima, Takeshi Ohwaki, Yasunori Taga, Effect of H$_2$O partial pressure and temperature during Ti sput-

tering on texture and electromigration in AlSiCu/Ti/TiN/Ti metallization, *J. Vac. Sci. Technol B*, Vol. **16** (5) p. 2751 (1998).

R. W. Ryan, R. F. Kopf, R. A. Hamm, R. J. Malik, R. Masaitis, R. Opila, Dielectric-assisted trilayer lift-off process for improved metal definition, *J. Vac. Sci. Technol B*, Vol. **16** (5) p. 2759 (1998).

H. C. Lee, M. Creusen, S. Vanhaelemeersch, Feasibility of gate patterning by using a hard mask on 0.25 lm technology and below, *J. Vac. Sci. Technol B*, Vol. **16** (5) p. 2763 (1998).

Masahiro Hashimoto, Toshishige Koreeda, Nobuyoshi Koshida, Masanori Komuro, Nobufumi Atoda, Application of dual-functional MoO$_3$/WO$_3$ bilayer resists to focused ion beam nanolithography, *J. Vac. Sci. Technol B*, Vol. **16** (5) p. 2767 (1998).

H. Noguchi, Y. Kubota, I. Okada, M. Oda, T. Matsuda, A. Motoyoshi, S. Ohki, H. Yoshihara, Fabrication of x-ray mask from a diamond membrane and its evaluation, *J. Vac. Sci. Technol B*, Vol. **16** (5) p. 2772 (1998).

T. Ohmi, M. Yoshida, Y. Matudaira, Y. Shirai, O. Nakamura, M. Gozyuki, Y. Hashimoto, Development of a stainless steel tube resistant to corrosive Cl$_2$ gas for use in semiconductor manufacturing, *J. Vac. Sci. Technol B*,Vol. **16** (5) p. 2789 (1998).

Jeong Sook Ha, Kang-Ho Park, Wan Soo Yun, El-Hang Lee, Nanometer scale selective etching of Si(111) surface using silicon nitride islands, *J. Vac. Sci. Technol B*, Vol. **16** (5) p. 2806 (1998).

M. R. Rakhshandehroo, J. W. Weigold, W.-C. Tian, S. W. Pang, Dry etching of Si field emitters and high aspect ratio resonators using an inductively coupled plasma source, *J. Vac. Sci. Technol B*, Vol. **16** (5) p. 2849 (1998).

Xiaomeng Chen, Harry L. Frisch, Alain E. Kaloyeros, Barry Arkles, Low temperature plasma-promoted chemical vapor deposition of tantalum from tantalum pentabromide for copper metallization, *J. Vac. Sci. Technol B*, Vol. **16** (5) p. 2887 (1998).

Sang Won Kang, Jae Soo Yoo, Jong Duk Lee, Photolithographic patterning of phosphor screens by electrophoretic deposition for field emission display application, *J. Vac. Sci. Technol B*, Vol. **16** (5) p. 2891 (1998).

MATERIALS SCIENCE & ENGINEERING B

T. Ami, M. Suzuki, MOCVD growth of (100)-oriented CeO$_2$ thin films on hydrogen-terminated Si(100) substrates, *Mat. Sci. & Eng. B*, Vol **54** (1-2) p. 84 (1998).

R. Schmid-Fetzer, R. Wenzel, F. Goesmann, Bulk diffusion studies of metallizations on titanium- based contacts at 600°C, *Mat. Sci. & Eng. B*, Vol **54** (1-2) p.175 (1998).

SOLID-STATE ELECTRONICS

H. Fukutome, K. Takano, S. Hasegawa, H. Nakashima, T. Aoyama, H. Arimoto, Direct imaging of nano pn junctions and their bulk electronic properties with the use of scanning tunneling microscopy, *Solid-State Elec.*, vol **42** (7-8) p. 1075(1998).

K. Komori, F. Sasaki, X. Wang, M. Ogura, H. Matsuhata, Barrier thickness dependence of optical properties in GaAs coupled quantum wires, *Solid-State Elec.*, vol **42** (7-8) p. 1211 (1998).

Y. Hanada, N. Ono, H. Fujikura, H. Hasegawa, Direct formation of InGaAs coupled quantum wire-dot structures by selective molecular beam epitaxy on InP patterned substrates, *Solid-State Elec.*, vol **42** (7-8) p. 1413 (1998).

M. Inada, T. Kikutani, H. Hori, S. Yamada, Fabrication of GaAs-Ge-GaAs lateral narrow junctions and low-temperature hole transport, *Solid-State Elec.*, vol **42** (7-8) p. 1539 (1998).

B. Shim, S. Torii, T. Ota, K. Kobayashi, K. Maehashi, S. Hasegawa, K. Inoue, H. Nakashima, Formation of InGaAs strained quantum wires on GaAs vicinal (110) substrates grown by molecular beam epitaxy, *Solid-State Elec.*, vol **42** (7-8) p.1609 (1998).

T. Mukaiyama, K. Saito, H. Ishikuro, M. Takamiya, T. Saraya, T. Hiramoto, Fabrication of gate-all-around MOSFET by silicon anisotropic etching technique, *Solid-State Elec.*, vol **42** (7-8) p.1623 (1998).

M. Watanabe, W. Saitoh, Y. Aoki, J. Nishiyama, Epitaxial growth of nanometer-thick CaF_2/CdF_2 heterostructures using partially ionized beam epitaxy, *Solid-State Elec.*, vol **42** (7-8) p.1627 (1998).

L. Lai, Y. Chan, Selectively dry-etched n + -GaAs/AlGaAs/n-InGaAs doped-channel FETs by using a $CHF_3 + BCl_3$ plasma, *Solid-State Elec.*, vol **42** (10) p. 1793 (1998).

C. W. Kuo, Y. K. Su, H. H. Lin, C. Y. Tsia, Study and application of reactive ion etching on GaInP/InGaAs/GaInP quantum well HEMTs, *Solid-State Elec.*, vol **42** (11) p. 1933 (1998).

J. A. Diniz, J. W. Swart, K. B. Jung, J. Hong, S. J. Pearton, Inductively coupled plasma etching of In-based compound semiconductors in $CH_4/H_2/Ar$, *Solid-State Elec.*, vol **42** (11) p. 1947 (1998).

R. J. Shul, F. Ren, E. Wolfgang, Power semiconductor devices and processes, *Solid-State Elec.*, vol **42** (12) p. 2117 (1998).

R. J. Shul, G. A. Vawter, C. G. Willison, M. M. Bridges, J. W. Lee, S. J. Pearton, C. R. Abernathy, Comparison of plasma etch techniques for III-V nitrides, *Solid-State Elec.*, vol **42** (12) p. 2259 (1998).

R. J. Shul, C. G. Willison, M. M. Bridges, J. Han, J. W. Lee, S. J. Pearton, C. R. Abernathy, J. D. Mackenzie, S. M. Donovan, High-density plasma etch selectivity for the III-V nitrides, *Solid-State Elec.*, vol **42** (12) p. 2269 (1998).

H. Cho, C. B. Vartuli, C. R. Abernathy, S. M. Donovan, S. J. Pearton, R. J. Shul, J. Han, Cl_2-based dry etching of the AlGaIn system in inductively coupled plasmas, *Solid-State Elec.*, vol **42** (12) p. 2277 (1998).

J. J. Wang, E. S. Lambers, S. J. Pearton, M. Ostling, C. M. Zetterling, J. M. Grow, F. Ren, R. J. Shul, ICP etching of SiC, *Solid-State Elec.*, vol **42** (12) p. 2283 (1998).

M. Fu, V. Sarvepalli, R. K. Singh, C. R. Abernathy, X. Cao, S. J. Pearton, J. A. Sekhar, A novel technique for RTP annealing of compound semiconductors, *Solid-State Elec.*, vol **42** (12) p. 2335 (1998).

THIN SOLID FILMS

P. A. Stampe, R. J. Kennedy, Growth of MgO on Si(100) and GaAs(100) by laser ablation, *Thin Solid Films*, vol **326** (1-2) p. 63 (1998)

PERGAMON

Materials Science in Semiconductor Processing 1 (1998) 343–350

MATERIALS
SCIENCE IN
SEMICONDUCTOR
PROCESSING

Patents ALERT

This section contains abstracts of recently issued patents in the United States and published patent applications filed from over 90 countries under the Patent Cooperation Treaty and compiled in accordance with interest profiles developed by the Editors.

Further information about complete patents can be obtained from:

REEDFAX Document Delivery System

275 Gibraltar Road, Horsham, PA 19044, USA

Phone: +1 215 441-4768

Fax:+1 215 441-5463

who offer a 24-hour, 7-days a week service.

See overleaf for details of a special service for readers of

Materials Science in Semiconductor Processing

An offer from

DOCUMENTS ON DEMAND ANYTIME, ANYWHERE

Free to readers of: *Materials Science in Semiconductor Processing*

REEDFAX™ Document Delivery System FAX FORM

FAX TODAY (our local international access code) **+1-215-441-5463**

As a first-time user, you are eligible to receive 5 FREE U.S. patents *(you pay only for the cost of delivery)*. Complete and return to confirm your personal 24-hour REEDFAX™ account number.

Yes! By return fax, please confirm my personal 24-hour REEDFAX account number and put me on-line — without cost or obligation — for **5 FREE United States patents** *(excluding delivery charges)*.

AUTHORIZED SIGNATURE _____

NAME (please print) _____

TITLE _____

COMPANY _____

ADDRESS _____

CITY _____ STATE _____ ZIP _____

PHONE _____ FAX _____ Alternate FAX (if desired) _____

Name to appear on faxed patent documents. If same as above, check box: ☐

NAME (please print) _____

Check here if you plan to use Client Numbers (Client charge-back numbers) ☐

Tax exempt number (For PA, FL and NY) _____

Name and address for billing. If same as above, check box: ☐

_____

_____

Please list your other locations or other individuals who would be interested in REEDFAX Document Delivery System. Use additional sheets if necessary.

_____

_____

5736457

METHOD OF MAKING A DAMASCENE METALLIZATION

Zhao Bin   Austin, TX, UNITED STATES assigned to Sematech

A semiconductor process and structure is provided for use in single or dual damascene metallization processes. A thin metal layer which serves as an etch stop and masking layer is deposited upon a first dielectric layer. Then, a second dielectric layer is deposited upon the thin metallization masking layer. The thin metallization masking layer provides an etch stop to form the bottom of the in-laid conductor grooves. In a dual damascene process, the thin metallization masking layer leaves open the via regions. Thus, the conductor grooves above the metallization masking layer and the via regions may be etched in the first and second dielectrics in one step. In a single damascene process, the thin metallization etch masking layer may cover the via regions. The etch stop and masking layer can be formed from any conductive or non-conductive materials whose chemical, mechanical, thermal and electrical properties are compatible with the process and circuit performance.

5736743

METHOD AND APPARATUS FOR ION BEAM FORMATION IN AN ION IMPLANTER

Benveniste Victor M   Gloucester, MA, UNITED STATES assigned to Eaton Corporation

A low energy ion implanter having an ion source for emitting ions and an implantation chamber spaced from the ion source by an ion beam path through which ions move from the source to the implantation chamber. A mass analyzing magnet positioned along the beam path between the source and the implantation chamber deflects ions through controlled arcuate paths to filter ions from the beam while allowing certain other ions to enter the ion implantation chamber. The magnet includes multiple magnet pole pieces constructed from a ferromagnetic material and having inwardly facing pole surfaces that bound at least a portion of a ion

deflection region. One or more current carrying coils set up dipole magnetic fields in the deflection region near the pole pieces. Additional coils help set up a quadrapole field in deflection region. A controller electrically coupled to the one or more coils of said magnet for controls current through the one or more current carrying coils to create the magnetic field in the deflection region near the pole pieces.

5736772

BIFURCATED POLYSILICON GATE ELECTRODES AND FABRICATION METHODS

Ko Young-Wi; Cho Yun-Jin; Cho Sung-Hee; Lee Hyong-Gon   Seoul, KOREA assigned to Samsung Electronics Co Ltd

A polysilicon gate electrode of an integrated circuit field effect transistor is formed in two portions which are isolated from one another. The first portion is formed on the gate insulating region. The second portion is formed on the semiconductor substrate outside the gate insulating region and is electrically insulated from the first portion. Since the first and second portions of the polysilicon gate electrode are isolated from one another, only the charge which is on the first polysilicon portion contributes to gate insulating region degradation during plasma etching. After polysilicon gate electrode formation, the first and second portions may be electrically connected by a link. Field effect transistor performance and/or reliability are thereby increased.

5737250

METHOD AND SYSTEM FOR SIMULATING ION IMPLANTATION FOR PERFORMING SIMULATION WITH AVOIDING OVERFLOW BY ADJUSTING MEMORY CONSUMING AMOUNT

Sawahata Koichi   Tokyo, JAPAN assigned to NEC Corporation

An ion implantation simulation system includes a grid generating portion for generating an orthogonal grid with respect to a two-dimensional configuration of a simulation object, an elongated segment extracting portion for extracting elongated segments defined by two grid lines in the orthogonal grid, a cell analyzing portion for extracting cells defined by adjacent grid lines perpendicular to the longer edge of the elongated segment, in the extracted elongated segments, and linearly rearranging polygon elements presenting in the cell along the longer edge direction, simulation performing portion performing linear ion implantation simulation with respect to the cell, in which the polygon elements are linearly rearranged, and a calculation result registering portion for registering an impurity concentration obtained as a result of simulation for each polygon element and registering the impurity concentration in each polygon element.

5738562

APPARATUS AND METHOD FOR PLANAR END-POINT DETECTION DURING CHEMICAL-MECHANICAL POLISHING

Doan Trung Tri; Sandhu Gurtej Singh; Grief Malcolm K Boise, ID, UNITED STATES assigned to Micron Technology Inc

A chemical-mechanical polishing apparatus includes a slurry-wetted polishing pad attached to a substantially planar surface of a platen. A wafer carrier is positioned in close proximity to the platen, and it has a substantially planar surface to which one side of a semiconductor wafer is removably attachable so that an opposing side of the semiconductor wafer is disposed against the polishing pad. An actuator imparts a translational motion to the platen so that the polishing pad moves relative to and in polishing contact with the semiconductor wafer. A sensor detects a change in the imparted translational motion corresponding to a change in the coefficient of friction between the polishing pad and the opposing side of the semiconductor wafer indicative of a planar end point on the opposing side of the semiconductor wafer. The sensor preferably includes a laser and a laser detector using a laser reflection or laser interferometric method to detect the change in the imparted translational motion. Also, the apparatus preferably includes a controller coupled to the sensor and the actuator to adjust the

actuator in response to the sensor detecting a change in the imparted translational motion.

5738574

CONTINUOUS PROCESSING SYSTEM FOR CHEMICAL MECHANICAL POLISHING

Tolles Robert D; Shendon Nor; Somekh Sasson; Perlov Ilya; Gantvarg Eugene; Lee Harry Q Santa Clara, CA, UNITED STATES assigned to Applied Materials Inc

An apparatus for polishing semiconductor wafers and other workpieces that includes polishing pads mounted on respective platens at multiple polishing stations. Multiple wafer heads, at least one greater in number than the number of polishing stations, can be loaded with individual wafers. The wafer heads are suspended from a carousel, which provides circumferential positioning of the heads relative to the polishing pads, and the wafer heads oscillate radially as supported by the carousel to sweep linearly across the respective pads in radial directions with respect to the rotatable carousel. Each polishing station includes a pad conditioner to recondition the polishing pad so that it retains a high polishing rate. Washing stations may be disposed between polishing stations and between the polishing stations and a transfer and washing station to wash the wafer as the carousel moves. A transfer and washing station is disposed similarly to the polishing pads. The carousel simultaneously positions one of the heads over the transfer and washing station while the remaining heads are located over polishing stations for wafer polishing so that loading and unloading of wafers and washing of wafers and wafer heads can be performed concurrently with wafer polishing. A robot positioned to the side of the polishing apparatus automatically moves cassettes filled with wafers into a holding tub, and transfers individual wafers vertically held in the cassettes between the holding tub and the transfer and washing station. The multiple polishing pads can be used to sequentially polish a wafer held in a wafer head in a step of multiple steps. The steps may be equivalent, may provide polishes of different finish, or may be directed to polishing different levels.

5739554

DOUBLE HETEROJUNCTION LIGHT EMITTING DIODE WITH GALLIUM NITRIDE ACTIVE LAYER

Edmond John A; Kong Hua-Shuang Cary, NC, UNITED STATES assigned to Cree Research Inc

A double heterostructure for a light emitting diode comprises a layer of aluminum gallium nitride having a first conductivity type; a layer of aluminum gallium nitride having the opposite conductivity type; and an active layer of gallium nitride between the aluminum gallium nitride layers, in which the gallium nitride layer is co-doped with both a Group II acceptor and a Group IV donor, with one of the dopants being present in an amount sufficient to give the gallium nitride layer a net conductivity type, so that the active layer forms a p-n junction with the adjacent layer of aluminum gallium nitride having the opposite conductivity type.

5741070

APPARATUS FOR REAL-TIME SEMICONDUCTOR WAFER TEMPERATURE MEASUREMENT BASED ON A SURFACE ROUGHNESS CHARACTERISTIC OF THE WAFER

Moslehi Mehrdad Mahmud Dallas, TX, UNITED STATES assigned to Texas Instruments Incorporated

A sensor (100) for measuring semiconductor wafer (10) temperature in semiconductor processing equipment (30), comprising a first laser (104) to provide a first laser beam at a first wavelength and a second laser (106) to provide a second laser beam at a second wavelength. The sensor also includes a laser driver (108) and an oscillator (110) to modulate the wavelength of the first and second laser beams as the laser beams are directed to and reflected from the wafer (10), and a detector module (130) to measure the change in specular reflectance of the wafer (10) resulting from the modulation of the wavelength of the first and second laser beams. The sensor system also includes signal processing circuitry (138) to determine rms surface roughness of the wafer (10) at a known reference temperature from the change in reflectance of wafer (10) resulting from modulation of the wavelengths of the first and second laser beams, and to determine the temperature of wafer (10) from the change in specular reflectance of wafer (10) resulting from modulation of the wavelengths of the first and second laser beams while wafer (10) is at an unknown temperature and the surface roughness of the wafer at the known temperature.

5741614

ATOMIC FORCE MICROSCOPE MEASUREMENT PROCESS FOR DENSE PHOTORESIST PATTERNS

McCoy John H; Suwa Kyoichi San Carlos, CA, UNITED STATES assigned to Nikon Corporation

Accurate measurement of the sidewalls of photoresist features formed on a semiconductor substrate is achieved by a double mask exposure process. This allows probing the sidewalls of closely spaced photoresist features with the probe tip of an atomic force microscope, in spite of the small (submicron) physical dimensions involved. First a conventional line/space pattern is exposed onto the photoresist using the desired mask. Then a second exposure is made using a second mask which has a special space pattern to effectively remove the already exposed photoresist features along at least one side of one of the previously exposed features. Hence, at least that one side of that one feature is clear of any adjoining photoresist features when the photoresist is then developed after the two exposures. This allows easy access to the sidewall of that one photoresist feature by tilting the probe tip of the atomic force microscope. This allows the measurement of the photoresist feature sidewall characteristics, including for instance angle, curvature and any artifacts present.

5741736

PROCESS FOR FORMING A TRANSISTOR WITH A NONUNIFORMLY DOPED CHANNEL

Orlowski Marius K; Baker Frank Kelsey   Austin, TX, UNITED STATES assigned to Motorola Inc

A semiconductor device (83)including a transistor (85) with a nonuniformly doped channel region can be formed with a relatively simple process without having to use high dose implants or additional heat cycles. In one embodiment, a polysilicon layer (14) and silicon nitride layer (16) are patterned at the minimum resolution limit. The polysilicon layer is then isotropically etched to form a winged gate structure (32). A selective channel implant step is performed where ions are implanted through at least one of the nitride wings of the winged gate structure (32) but are not implanted through the polysilicon layer (14). Another polysilicon layer (64)is conformally deposited and etched such that the polysilicon (74) does not extend beyond the edges of the nitride wings.

5742176

METHOD OF MEASURING A FE-B CONCENTRATION OF A SILICON WAFER

Kato Hirotaka; Matsumoto Kei Hiratsuka, JAPAN assigned to Komatsu Electronic Metals Co Ltd

An evaluation method of a silicon wafer by correctly calculating the Fe-B concentration is disclosed. Even when the SPV method is utilized, the over-estimated Fe-B concentration in silicon wafers containing oxygen-precipitation defects can be avoided. Diffusion lengths Lb and La of minority carriers in a P-type silicon wafer before and after an activation step are measured by the SPV method. A value of (Lb-La)/Lb calculated from La and Lb is compared with a constant C which is read from the plot of Lb vs. (Lb-La). If (Lb-La)/Lb is smaller than constant C, the concentration calculation is terminated since there are oxygen-precipitation defects in the silicon wafer. The calculation is car-

ried out for silicon wafers containing no oxygen-precipitation defects, and is based on the formula of Fe-B concentration (cm-3) approx $= 1*1016(La-2-Lb-2)$. Therefore, the Fe-B concentration can be precisely determined even though the silicon wafers in which a high-density of oxygen-precipitation defects exist are mixed together with silicon rods.

5744192

METHOD OF USING WATER VAPOR TO INCREASE THE CONDUCTIVITY OF COOPER DESPOSITED WITH CU(HFAC)TMVS

Nguyen Tue; Senzaki Yoshihide; Kobayashi Masato; Charneski Lawrence J; Hsu Sheng Teng Vancouver, WA, UNITED STATES assigned to Sharp Microelectronics Technology Inc; Sharp Kabushiki Kais

A method of blending water vapor with volatile Cu(hfac)TMVS (copper hexafluoroacetylacetonate trimethylvinylsilane) is provided which improves the deposition rate of Cu, without degrading the resistivity of the Cu deposited upon an integrated circuit surface. The method of the present invention uses a relatively small amount of water vapor, approximately 0.3 to 3% of the total pressure of the system in which chemical vapor deposition (CVD) Cu is applied. The method specifies the flow rates of the liquid precursor, carrier gas, and liquid water. The method also specifies the pressures of the vaporized precursor, vaporized precursor blend including carrier gas and water vapor. In addition, the temperatures of the vaporizers, chamber walls, and IC surfaces are disclosed. A Cu precursor blend is also provided comprising vaporized Cu(hfac)TMVS and water vapor. The ratio of water vapor pressure to vaporized precursor is approximately 0.5 to 5%. Further, an IC surface covered with Cu applied with a Cu precursor blend including vaporized Cu(hfac)TMVS and water vapor, with the above mentioned ratio of water vapor pressure to volatile Cu(hfac)TMVS pressure, is provided.

5744839

ESD PROTECTION USING SELECTIVE SILICIDING TECHNIQUES

Ma Manny K F; Schoenfeld Aaron Boise, ID, UNITED STATES assigned to Micron Technology Inc

The present invention relates to methods and apparatus for manufacturing semiconductor devices, and in particular for forming electrostatic discharge (ESD) protection devices, using selective siliciding, in a CMOS integrated circuit. Predetermined discharge paths are created for discharging input and output buffer pads, during an ESD event, through ESD protection devices. During fabrication, an oxide layer is utilized as a mask to prevent silicided regions from forming in source/drain regions, self-aligned with the gates. The buffer transistor gate-to-contact spacing is made longer than the gate-to-contact spacing in the associated protection transistor, to shunt charge through the protection device. In a further embodiment, active area resistance is formed between the output/input buffer transistor and the ESD protection device, to further increase the resistance of the path between the buffer pad to the associated buffer transistor.

5747830

SEMICONDUCTOR DISPLAY DEVICE WITH A HYDROGEN SUPPLY AND HYDROGEN DIFFUSION BARRIER LAYERS

Okita Akira Yamato, JAPAN assigned to Canon Kabushiki Kaisha

The invention provides a semiconductor device having a low threshold voltage and capable of operating at a high speed, and also provides an active matrix display device including such a semiconductor device. The invention also provides a method of producing such a semiconductor device and an active matrix display device. In the invention, a hydrogen supply layer is disposed above semiconductor layers in such a manner that the hydrogen supply layer is apart from the semiconductor layers. A hydrogen diffusion barrier layer for preventing hydrogen from diffusing outward from the hydrogen supply layer is disposed on the hydrogen supply layer in such a manner that the hydrogen diffusion barrier layer is in direct contact with the hydrogen supply layer. The hydrogen diffusion barrier layer is preferably made of a high melting point metal or a compound thereof. The hydrogen supply layer is preferably formed by depositing SiN or amorphous silicon by means of plasma CVD.

Printed and bound by CPI Group (UK) Ltd, Croydon, CR0 4YY

14/05/2025

01871070-0001